T0269249

The algebraic theory of modular systems

The algebraic theory of modular systems

F. S. Macaulay

with a new Introduction by

PAUL ROBERTS

University of Utah

CAMBRIDGE
UNIVERSITY PRESS

CAMBRIDGE
UNIVERSITY PRESS

University Printing House, Cambridge CB2 8BS, United Kingdom

Cambridge University Press is part of the University of Cambridge.

It furthers the University's mission by disseminating knowledge in the pursuit of education, learning and research at the highest international levels of excellence.

www.cambridge.org
Information on this title: www.cambridge.org/9780521455626

Copyright Cambridge University Press 1916
Introduction © Cambridge University Press 1994

This publication is in copyright. Subject to statutory exception and to the provisions of relevant collective licensing agreements, no reproduction of any part may take place without the written permission of Cambridge University Press.

First published 1916
Reissued with an Introduction by Paul Roberts
in the Cambridge Mathematical Library 1994

A catalogue record for this publication is available from the British Library

ISBN 978-0-521-45562-6 Paperback

Cambridge University Press has no responsibility for the persistence or accuracy of URLs for external or third-party internet websites referred to in this publication, and does not guarantee that any content on such websites is, or will remain, accurate or appropriate.

CONTENTS

III. GENERAL PROPERTIES OF MODULES

IV. THE INVERSE SYSTEM

DEFINITIONS

LIST OF REFERENCES

References to the following in the text are given by their initial letters, e.g., (L, p. 51).

(BN) A. Brill and M. Noether, "Ueber die algebraischen Functionen und ihre Anwendung in der Geometrie" (*Math. Ann.* 7 (1874), p. 269).

(D) R. Dedekind, "Sur la Théorie des Nombres entiers algébriques" (*Bull. Sc. Math.* (1) 11 (1876), p. 278/288; (2) 1 (1877), p. 17/41, 69/92, 144/164, 207/248). Also in G. Lejeune Dirichlet's *Vorlesungen über Zahlentheorie* (Brunswick, 2nd edition (1871) and 4th edition (1894)).

(DW) R. Dedekind and H. Weber, "Theorie der algebraischen Functionen einer Veränderlichen" (*J. reine angew. Math.* 92 (1882), p. 181).

(E) E. B. Elliott, *Algebra of Quantics* (Clarendon Press, 2nd ed. 1913).

(H) D. Hilbert, "Ueber die Theorie der algebraischen Formen" (*Math. Ann.* 36 (1890), p. 473).

(H$_1$) D. Hilbert, "Ein allgemeines Theorem über algebraische Formen" (*Math. Ann.* 42 (1893), p. 320), §3 of "Ueber die vollen Invariantensysteme."

(K) J. König, *Einleitung in die allgemeine Theorie der algebraischen Grössen* (B. G. Teubner, Leipzig, 1903).

(Kr) L. Kronecker, (i) "Grundzüge einer arithmetischen Theorie der algebraischen Grössen" (*J. reine angew. Math.* 92 (1882), p. 1). Also expounded with additions in (K) especially in Chap. IX.

(ii) "Zur Theorie der Formen höherer Stufen" (*Sitz. Akad. d. Wiss. zu Berlin*, 37 (1883), p. 957).

(L) E. Lasker, "Zur Theorie der Moduln und Ideale" (*Math. Ann.* 60 (1905), p. 20).

(M) F. S. Macaulay, "On the Resolution of a given Modular System into Primary Systems" (*Math. Ann.* 74 (1913), p. 66).

(M$_1$) F. S. Macaulay, "The Theorem of Residuation" (*Proc. Lond. Math. Soc.* (1) 31 (1900), p. 381).

(M$_2$) F. S. Macaulay, "Some Formulae in Elimination" (*Proc. Lond. Math. Soc.* (1) 35 (1903), p. 3).

(M₃) F. S. Macaulay, "On a Method of dealing with the Intersections of Plane Curves" (*Trans. Am. Math. Soc.* 5 (1904), p. 385).

(Mo) E. H. Moore, "The decomposition of modular systems of rank n in n variables" (*Bull. Am. Math. Soc.* (2) 3 (1897), p. 372).

(N) M. Noether, "Ueber einen Satz aus der Theorie der algebraischen Functionen" (*Math. Ann.* 6 (1873), p. 351).

(Ne) E. Netto, "Zur Theorie der Elimination" (*Acta Math.* 7 (1886), p. 101).

(S) C. A. Scott, "On a Method for dealing with the Intersections of Plane Curves" (*Trans. Am. Math. Soc.* 3 (1902), p. 216).

(Sa) G. Salmon, "On the order of Restricted Systems of Equations' (*Modern Higher Algebra*, Dublin, 4th ed. (1885), Lesson xix).

(W₁) *Encyklopädie der Mathematischen Wissenschaften* (Teubner, Leipzig, Teil I, Bd. I, Heft 3 (1899), p. 283). G. Landsberg, "Algebraische Gebilde, etc."

(W₂) *Encyclopédie des Sciences Mathématiques* (Gauthier-Villars, Paris, Tome I, Vol. 2, Fasc. 2, 3 (1910, 11), p. 233). J. Hadamard and J. Kürschak, "Propriétés générales des corps, etc."

This account of the theory is founded upon that of G. Landsberg but is much fuller both in snbject matter and references.

PREFACE

THE present state of our knowledge of the properties of Modular Systems is chiefly due to the fundamental theorems and processes of L. Kronecker, M. Noether, D. Hilbert, and E. Lasker, and above all to J. König's profound exposition and numerous extensions of Kronecker's theory (p. xi). König's treatise might be regarded as in some measure complete if it were admitted that a problem is finished with when its solution has been reduced to a finite number of feasible operations. If however the operations are too numerous or too involved to be carried out in practice the solution is only a theoretical one; and its importance then lies not in itself, but in the theorems with which it is associated and to which it leads. Such a theoretical solution must be regarded as a preliminary and not the final stage in the consideration of the problem.

In the following presentment of the subject Section I is devoted to the Resultant, the case of n equations being treated in a parallel manner to that of two equations; Section II contains an account of Kronecker's theory of the Resolvent, following mainly the lines of König's exposition; Section III, on general properties, is closely allied to Lasker's memoir and Dedekind's theory of Ideals; and Section IV is an extension of Lasker's results founded on the methods originated by Noether. The additions to the theory consist of one or two isolated theorems (especially §§ 50—53 and § 79 and its consequences) and the introduction of the Inverse System in Section IV.

The subject is full of pitfalls. I have pointed out some mistakes made by others, but have no doubt that I have made new ones. It may be expected that any errors will be discovered and eliminated in due course, since proofs or references are given for all major and most minor statements.

I take this opportunity of thanking the Editors for their acceptance of this tract and the Syndics of the University Press for publishing it.

F. S. MACAULAY.

LONDON,
   June, 1916.

Introduction

Many of the ideas introduced in Macaulay's book on Modular Systems have developed into central concepts in the branch of mathematics known today as Commutative Algebra. His name is remembered today mostly through the term "Cohen–Macaulay ring", a notion which grew out of the unmixedness theorem in the third chapter of this book. However, it is less well known that he pioneered several other fundamental ideas, including the concept of Gorenstein ring and the use of injective modules, ideas which were not systematically developed until considerably later in this century.

In 1916, when Macaulay's book appeared, the field of Commutative Algebra had not grown to the point where it could be considered a separate branch of Mathematics, and rings were not studied for their own sake. The topics in this book had their origins instead in the problem of finding solutions to systems of polynomial equations. This problem may be considered to be a branch of Algebraic Geometry, and many of the subjects discussed here really belong to that field. While it is not always easy (or necessary) to separate the field of Commutative Algebra from the more algebraic side of Algebraic Geometry, we concentrate here on developments in Commutative Algebra, since the main new methods introduced in this book would today be considered as belonging to that field.

This introduction has several aims. We describe many of the ideas in this book, both in their own context and how they have developed since those days. We first present a summary of how the approaches differed, and then give a more detailed account of how the individual ideas have developed. Macaulay's writing is not always easy to read, in part because of the condensed style, and in part because of the differences in terminology and notation, not to mention certain conventions which are no longer in common use. For example, the word "module" is used where we would now use the word "ideal"; while "module" has a very different meaning today. We point out some of these differences in the introduction, and at the end we summarize the main differences between his terminology and that which is currently standard. We use modern terminology unless stated otherwise. A polynomial ring in n variables will be denoted $k[x_1,\ldots,x_n]$, $k[x_i]$, or R, and an ideal in this ring will be denoted M.

One obvious difference between the content of this book and that of a modern text in the subject is that the modern theory is much more abstract, while that of Macaulay involves more specific computations. In fact, one of the reasons that Macaulay's work has been referred to steadily over the years is that it contains many useful examples which are still of widespread interest.

Background: the study of polynomial rings in Macaulay's time.

The first two chapters of "The Algebraic Theory of Modular Systems" deal mostly with the question of describing solutions to sets of polynomial equations, while the last two are concerned with the finer structure of the ideals themselves. We begin by giving a picture of the subject of describing these solutions and how it fit into the development of mathematics at the time.

The ideal solution to the question of finding all solutions to a set of equations would be to list variables which can be assigned values arbitrarily and provide a set of formulas for each of the other variables in terms of the arbitrary ones. In simple cases, such as when the equations are linear, such a program can be carried out, but such complete solutions cannot be found in general. The first two chapters of this book deal with partial solutions to this problem.

While one cannot "solve" a set of equations as above, much can be said about the set of solutions. One main fact, known well before Macaulay's time, is that the set of solutions can be divided into components, each of which is "irreducible" and has a well-defined dimension. A further major step was Hilbert's Nullstellensatz [7], in which he showed that there is a correspondence between certain ideals of the ring of polynomials and sets of solutions; on the other hand, the ideal structure was finer in that different ideals may have the same set of solutions. The work of Lasker [12] on primary decomposition showed that an ideal can also be divided into components, or, more precisely, that every ideal may be represented as an intersection of primary ideals. Certain of these primary ideals correspond to the components of the set of solutions, but in general there are others, which are called imbedded components. It was Macaulay who pointed out the importance of these imbedded components and their relation to the structure of the ideal. We discuss these questions further in the section on the third chapter of the book.

The fact that the subject was still close to the problem of finding solutions to polynomial equations may explain one convention which appears strange to a modern reader. Whereas it would seem normal today to say that an ideal M is contained in an ideal N if M is a subset of N, Macaulay would say that M contains N in this case. On the other hand, if we think of the ideals as representing their sets of solutions, Macaulay's expression is reasonable. It could be argued that the modern term for his "module" is "closed subscheme of affine space" rather than "ideal in a polynomial ring", since closed subschemes correspond to ideals with the order relation reversed. However, his arguments are ideal-theoretic, and we shall refer to them as ideals.

A comparison with the modern approach to the subject.

Since Macaulay's book appeared, the field of Commutative Algebra has changed considerably. The most obvious difference between the work of Macaulay and a modern book in Commutative Algebra is that Macaulay dealt exclusively with ideals in polynomial rings, while more recently the subject is much more abstract and deals with arbitrary "Noetherian rings". For a basic modern text on the subject we refer to *Commutative Ring Theory* by Matsumura [15]; most of the results and facts which we mention here without a particular reference may be found in Matsumura's book. We also note the historical treatments of the subject by Kaplansky [10] and Bourbaki [3]. This introduction is not intended to give a complete history of the subject, but rather to point the influences of Macaulay's work, as well as some of the ways in which the subject has changed.

Although this tendency toward abstraction has resulted in greater generality, with a much wider class of rings being studied, the case of polynomial rings is still the most important example, and much of the field is devoted to it today. In fact, polynomial rings have a more central place in the subject than they did a few years ago, partly because of the recent introduction of computer methods, which are almost entirely devoted to computations in polynomial rings. We note that Macaulay's name is also commemorated in the most widely used system at present, the program *Macaulay* of Bayer and Stillman; while the contents of this book are only indirectly related to modern computational methods, his later work (Macaulay [13]) has influenced this development.

In addition to the tendency toward abstraction, there have been several more specific changes which have served to clarify the main ideas of Commutative Algebra and to simplify many of the proofs. First was the introduction of the ascending chain condition on ideals, following the proof of Emmy Noether [16] of the existence of primary decomposition, greatly clarifying this theorem. Second was localization; in a modern treatment many, if not most, of the questions are reduced to the local case and many problems are simplified. In many places in this book, substitutes for localization are obtained through ad hoc methods. Finally, the concept of dimension, due to Krull [11], has been refined; it is no longer based on the complicated notions arising from explicit solutions to equations required in this book. We shall return these topics at greater length in further sections.

Chapters 1 and 2: The Resultant and The Resolvent.

As stated above, the first two chapters of the book are devoted to the questions of finding solutions to systems of polynomial equations, and much of the material here comes from earlier authors. Many of these topics have not been considered as essential to the subject recently as they were then, and for that reason Macaulay's book remains a good reference for this material.

The first chapter is devoted to the resultant of a set of n homogeneous polynomials in n variables. The aim of this technique is to find a polynomial in the coefficients of the n polynomials which vanishes if and only if the set of polynomials has a non-zero solution. This generalizes the well-known case of n linear homogeneous equations in n unknowns, which have a non-zero solution if and only if the determinant of the coefficient matrix vanishes. In the case of two polynomials in two variables, the resultant is again the determinant of a matrix whose entries are the coefficients; this case is still widely used today, particularly in the version of the resultant of two non-homogeneous polynomials in one variable, where it gives a criterion for the existence of a common solution to the polynomials. For polynomials in three or more variables, the resultant is defined as the greatest common divisor of a set of determinants, and is considerably more complicated. Macaulay gives a better method for constructing the resultant (§§6–7) and proves its basic properties (§§8–11), such as irreducibility and the fact that it does characterize sets of polynomials with non-zero solutions.

While most of this chapter studies the resultant for its own sake, we mention two connections with the material in later chapters. First, the resultant is defined via a matrix whose columns are indexed by the monomials in the polynomial ring and whose rows are indexed by monomial multiples of the n polynomials, so that the entries are either zeros or coefficients of the polynomials (see §1). This representation of elements of an ideal by an array of coefficients is used systematically in the fourth chapter to study more general ideals. Second, the resultant is referred to later (§67) in the discussion of multiplicities.

If the aim of the theory of the resultant is to characterize the existence of non-trivial solutions, the purpose of the resolvent is to find the solutions. In theory, the method succeeds; in practice, the computations are formidable, and they assume that every polynomial in one variable over any coefficient field can be solved. However, the resolvent, or rather the variation called the u-resolvent, is used in an important theoretical way in later chapters, so we describe this method briefly.

The basic idea here is to express the polynomials as polynomials in one variable x_n whose coefficients are polynomials in the $n-1$ variables $x_1, \ldots, x_{n-1}$, and to reduce the problem to one in fewer variables. We suppose first that the polynomials have a non-trivial common factor $f(x_i)$. In this case, the first $n-1$ variables can be chosen arbitrarily, and solving the polynomial $f(x_i)$ for x_n provides a solution to the complete set of equations. Since there are $n-1$ independent variables, the dimension of the space of solutions is $n-1$. The factor $f(x_i)$ is then factored out, and the resultant is then used to give a condition on the coefficients of the resulting polynomials (polynomials in $n-1$ variables) to have a solution, giving a new set of polynomials in one fewer variable (§§13–14). This process is continued, resulting in a (possibly trivial) factor at each stage. This factor is a polynomial in $n-k$ variables at the kth stage, and is constructed in the same way as the polynomial $f(x_i)$ of the first step. The product of these factors is the total resolvent.

Before discussing how Macaulay uses this construction we make one comment on the modern way of reducing the number of variables. Now one would take the intersection of the ideal with $k[x_1, \ldots, x_{n-1}]$ via standard bases rather than use the resultant, since this requires much less computation. While determinants are of important theoretical value, they are not efficient to compute, so other methods are used.

Each factor of the resolvent gives a polynomial whose solutions have a given dimension. In each factor, certain variables are taken to be arbitrary, and the equation can be solved for the other variables. Thus the factors give the irreducible components of the set of solutions of the polynomials together with a very concrete description of their dimension.

It is in this chapter (§17) that Macaulay introduces the discussion of unmixedness, which is among his most important contributions to the field. As stated above, the factors of the resolvent determine the irreducible components of the set of solutions. Macaulay calls this set of solutions the "spread" of the ideal. The question is whether the other factors, whose solutions are properly contained in one of the components, give a meaningful description of the finer structure of the ideal and a reasonable criterion for the ideal to be "mixed". Such a possibility had been suggested by earlier authors. Through a succession of examples, Macaulay shows that the factors of the resolvent are inadequate for determining whether an ideal is mixed or not. This discussion is taken up again in the third chapter.

The concept of dimension.

Macaulay's notion of dimension of a quotient R/M, where M is an ideal, is based on the "u-resolvent" of the ideal, which is a variation on the resolvent as described above. We refer to (§§18–22) for the definitions. The u-resolvent gives the solutions in a somewhat more precise form, and, like the ordinary resolvent, gives a very concrete interpretation of the components and the dimension of the set of solutions in the number of arbitrary variables. Macaulay's use of dimension is entirely based on this construction, and he uses the irreducible components defined in this way systematically in place of the corresponding prime ideals.

Since the work of Krull [11], it has been the prime ideals themselves which form the basis for dimension theory in Noetherian rings. The modern view is much simpler, in that it is not based on the resolvent, which is by any standards a complicated construction. The stronger results of dimension theory which apply to polynomial rings but not to general Noetherian rings are based on the Hilbert Nullstellensatz [7] rather than the resolvent. We note that Macaulay never mentions Hilbert's Nullstellensatz, although he does prove and use the Hilbert basis theorem in the third chapter.

If the introduction of Krull dimension made it possible to avoid the resolvent in dimension theory, the introduction of chain conditions by Emmy Noether [16] made the arguments using induction on dimension even simpler. A striking example is the proof of "Lasker's Theorem" (now attributed to Lasker and Noether) on primary decomposition (§39 in Chapter three). This proof is also based on the u-resolvent, which makes the induction part much more difficult than it needs to be; this step would now be obtained by simply taking a maximal counterexample and working from there. For the modern reader, these arguments can be avoided by using the modern version of dimension and chain conditions. This dimension theory is also at the basis of Macaulay's use of prime ideals, where, as mentioned above, he usually uses the corresponding set of solutions and its representation by the "u-resolvent" in place of the prime ideal itself. Replacing "irreducible spread" by "prime ideal" usually gives the same result and sometimes avoids complexities which are not totally necessary.

However, while the theory of the resolvent is no longer needed as a foundation for the theory of dimension and irreducible components, these sections of the book are still very interesting in their original purpose of describing solutions to sets of polynomial equations.

Chapter 3: General Properties of Modules.

The third chapter contains the results for which Macaulay is best known, including the definition of unmixed ideals and the Unmixedness Theorem for ideals in polynomial rings. It also contains many related theorems on unmixed ideals, as well as many other theorems which make up the basic theory of Commutative Algebra, including for example the existence of primary decomposition (discussed in the previous section) and the Hilbert basis theorem.

The first several sections of the chapter (§§23–28) are devoted to the basic constructions on ideals, and are essentially the same as would appear in a modern text except for notation. His notation is based on that of elementary number theory, while more recently the convention has been to use set-theoretic notation. For instance, he will refer to the LCM (least common multiple) of ideals where now we would call it the intersection, and he writes (M/M') for $M : M' = \{m | mM' \subseteq M\}$.

In the earlier chapters he had pointed out that the method of the resolvent was inadequate in differentiating between mixed and unmixed ideals. While this method gave a satisfactory theory of the components of the set of solutions corresponding to an ideal of polynomials, when there were imbedded components it may not detect them, while it produced extraneous components where they should not be any. The method which he uses to give a good definition of unmixedness is based on Lasker's theory of primary decomposition.

This theorem states that every ideal can be written as a finite intersection of primary ideals, and that the set of solutions corresponding to a primary ideal is irreducible, and thus corresponds to a unique prime ideal. The definitions of prime and primary ideals have not changed over the years, but, as mentioned above, Macaulay consistently refers to the "irreducible spread", or set of solutions, corresponding to a primary ideal rather than to the prime ideal itself. If the representation of an ideal M as an intersection of primary ideals is made irredundant, the set of prime ideals thus obtained is unique; he calls these the "relevant spreads" of the ideal M; they would now be called the associated prime ideals of R/M.

Macaulay's main contribution to this theory was to point out the importance of what he called 'imbedded spreads", now called imbedded primes of R/M. That is, among the relevant spreads of a module, it is possible that some are properly contained in others, and this will have consequences, particularly involving zero-divisors on the quotient. He now gives the correct definition of an unmixed ideal —it is one all of whose relevant spreads have the same dimension. In particular, an unmixed ideal can have no imbedded prime ideals.

We discuss this further below, but we wish to mention one idea which is not usually included among these theorems today. In the midst of the section on prime ideals, Macaulay introduces "H-modules", which means homogeneous ideals. For a given ideal, he then defines the concept of H-basis; this means a set of generators for the ideal whose leading coefficients generate the ideal of all leading coefficients of elements of the ideal. This concept is used in the more detailed theory of ideals in the fourth chapter. Its similarity to the concept of standard (Gröbner) basis, which has become important for modern computational methods, is quite evident. Standard bases have the same property as Macaulay's H-bases, but with respect to a total order on all monomials rather than the order by degree of the polynomials. In a later work ([13]) Macaulay studied these orders on monomials in studying Hilbert polynomials of ideals, and this paper has influenced modern developments in computational Commutative Algebra.

The Unmixedness Theorem and the concept of depth.

The main theorem of this chapter, and perhaps of the entire book, is the theorem that an ideal of height r generated by r elements is unmixed

(§§44–48). He calls such an ideal a "module of the principal class"; it would now usually be called a complete intersection ideal.

This theorem, in addition to being of great interest in itself, has had a tremendous influence on later developments in the subject. The concept of "depth" originates from this theorem; to explain this connection, it is useful to examine briefly Macaulay's proof of the theorem . Starting with an ideal of height r generated by r elements, he assumes that it has an associated prime ("relevant spread") of height greater than r; such a situation would violate the theorem. He then shows that one can find a polynomial of the form $x_i - a_i$ (perhaps after a change of co-ordinates) so that the new ideal generated by $r + 1$ elements has an associated prime of height at least $r + 2$. Continuing in this way, he eventually produces a complete intersection ideal of height less than n with a maximal associated prime ideal. In a separate Lemma (§44), he shows that this situation is impossible.

We now compare this proof to the modern idea of depth. The depth of a local ring can be defined as follows: if the maximal ideal is an associated prime ideal, the depth of the ring is zero. If not, there exists a non-zero-divisor in the maximal ideal of the ring, and one can divide by it. One can continue dividing by non-zero divisors until the maximal ideal is an associated prime ideal of the quotient. The number of steps, or the total number of successive non-zero-divisors found, is defined to be the depth of the ring. Such a sequence of non-zero-divisors is called a *regular sequence*. Thus Macaulay showed that the depth equals the dimension for a ring of the type R/M where M is generated by the number of elements equal to its height, and in the process he showed that if the depth is equal to the dimension, the ideal is unmixed.

This theorem was generalized to the case of a regular local ring some years later by Cohen [4]. Since then both of their names have been associated to this theorem, and rings which satisfy the theorem that an ideal of height r generated by r elements is unmixed have been called "Cohen–Macaulay rings". While Macaulay does not quite define this property, he does give an example to show that an unmixed ideal M does not necessarily have the property that $(M, x_i - a_i)$ is unmixed; this example is now the standard example of a non-Cohen–Macaulay integral domain. (§44).

While this concept has been used for some time to prove unmixedness of certain ideals, it was the introduction of homological algebra which led to an understanding of the true importance of this concept. For one thing, there are several equivalent definitions of depth based on the vanishing of homology modules (see Matsumura [15], chapter 6, for example) which make the theory simpler. Second, the Auslander–Buchsbaum theorem [1] gives an extremely useful relation between homological dimension and depth. This theorem states that if T is a module of finite projective dimension, we have

$$\text{depth}(T) + \text{proj.dim.}(T) = \text{depth}(R).$$

In addition, there are several vanishing theorems, such as Peskine and Szpiro's "Lemme d'acyclicité" [17] which say that if the depth of the ring is large enough, and the dimension of the homology of a complex is small enough, the complex is exact. These theorems have become the standard method for proving that a complex constructed as a free resolution of a module is in fact a resolution. As an example, the exactness of a resolution of an ideal generated by determinants is usually established in this way. While Macaulay never considers free resolutions in the modern sense, he has given an idea of this method in §§52 and 53, where he shows that the relations between the determinants are the predicted ones if the height of the ideal they generate is large enough. In addition, he shows that the ideal is unmixed in this case; these arguments may be considered to be precursors of the modern homological methods.

The concept of depth and the introduction of homological algebra have made it possible to prove many theorems relating depth, homological dimension, and multiplicities in the case of Cohen–Macaulay rings (while we have not yet mentioned multiplicities, they are discussed in the fourth chapter of the book). One of the main developments in the field of Commutative Algebra in the last twenty years or so has been has been to prove these in the case in which they were not Cohen–Macaulay, as well as many related conjectures. A major step was the introduction of the Frobenius map and reduction of these questions to positive characteristic by Peskine and Szpiro [17]. These techniques were further developed by Hochster [8], where a thorough discussion and list of these conjectures may be found. Recent progress on them is described in Roberts [19].

Chapter 4: The Inverse System.

The last chapter of the book is devoted to the "inverse system" of an ideal, which is one of the most original ideas in the book and one which Macaulay himself describes as "new". The basic idea of the inverse system is to study an ideal by investigating its dual, where the dual of an ideal is meant in a sense which will be described more precisely below. Using this idea, Macaulay studies deeper properties of ideals, including the concept of a principal system, which would now be called a Gorenstein ideal, and which has become another of the fundamental concepts in Commutative Algebra. In addition, it has led to very general duality theorems in Algebraic Geometry. Among the theorems in this chapter, he proves that a complete intersection is Gorenstein.

The construction of the inverse system.

The basic construction is fairly simple. In modern terms, it would be described as follows: corresponding to an ideal M in a polynomial ring $k[x_i]$, the inverse system is the dual of $k[x_i]/M$ over k, or $\text{Hom}_k(k[x_i]/M, k)$.

Macaulay's actual construction of the inverse system is considerably more concrete (§57). If M is an ideal of $k[x_i]$, the dual of $k[x_i]/M$ is naturally imbedded in the dual of $k[x_i]$. The dual of $k[x_i]$ is identified with a power series ring $k[[x_i^{-1}]]$. The monomials in x_i^{-1} are then dual to the corresponding monomials in x_i, and the natural structure of $k[x_i]$-module is ordinary multiplication, where we let any monomial in the product with at least one positive exponent equal 0. The duality pairing can be recovered as follows: if $f(x_i)$ is a polynomial and $g(x_i^{-1})$ an element of the inverse system, the value of $g(x_i^{-1})$ paired with $f(x_i)$ is the constant term of $f(x_i)g(x_i^{-1})$. The inverse system of M is then identified with the set of all power series which, when paired with all elements of M, give zero. If $g(x_i^{-1})$ is such an element, the equation $g(x_i^{-1}) = 0$ is called a "modular equation" of M.

As an example of this notation, which is used throughout the chapter, we note the rather disconcerting expression "$1 = 0$", which occurs several times. This equation should be interpreted to mean that "$1 = 0$" is a modular equation of M, which in turn means that the constant term of $f(x_i) \cdot 1$ is zero for all $f(x_i) \in M$. Thus, it means simply that M is contained in the ideal generated by $x_1, \ldots, x_n$.

As mentioned above, his construction of the inverse system of an ideal is very concrete, and it is based on a picture of the ideal which he calls the "dialytic array" (§59). The dialytic array corresponding to the ideal M is a list of all elements in a basis for M as a vector space over k arranged by degree. There are several versions of this construction, but in all of them one obtains an infinite matrix with rows indexed by elements of M and columns by monomials; this matrix is analogous to the matrix used to define the resultant in Chapter 1. The inverse system is then represented in a similar way with entries dual to those in the dialytic array of M.

We remark on one other convention which could be interpreted differently today, and that is that operations such as LCM defined on inverse systems mean the corresponding operation (in this case intersection) on the corresponding ideals, and not the intersection of the submodules of $k[x_i^{-1}]$, which could be inferred by a modern reader.

This construction has been transformed considerably over the years. In a modern context, one would consider a local ring A with residue field k, and the present version of the inverse system is the injective hull of the residue field of k, denoted $E(k)$. If one applies this to the localization of a polynomial ring at the maximal ideal generated by $x_1, \ldots, x_n$, we get a construction almost identical to that of Macaulay, but with polynomials in x_i^{-1} rather than arbitrary power series. It is interesting to note that Macaulay does single out the elements of the dual which have a finite number of non-zero coefficients, which he calls "Noetherian equations", and in effect uses this as a substitute for localization at the origin (§65). If M is an ideal of A, the submodule $\mathrm{Hom}(A/M, E(k))$ is the analogue of Macaulay's inverse system, and is usually called the Matlis dual of A/M; this terminology has been extended to A-modules, where it forms the basis for a duality theory between Noetherian and Artinian modules over a complete local ring (see Matlis [14]).

Uses of the Inverse System.

Next, we discuss how Macaulay applied the inverse system to the study of ideals. There are two parts to this discussion, the first dealing with ideals with quotients of finite length, and the second to more general unmixed ideals. The most significant property defined here is undoubtedly that of a principal system, which means an ideal whose dual is generated by one element. In the dimension zero case, this condition on an ideal M is precisely the condition for R/M to be a Gorenstein ring.

Macaulay first studies in detail the case in which $k[x_i]/M$ is a finite length ring supported at the origin, in which case he calls M a "simple module". (§ 67–75). The major result here is that an ideal of the "principal Noetherian class" (that is, an ideal whose localization at the origin is a complete intersection) is a principal system. In modern terms, this is the fundamental theorem that a local complete intersection is Gorenstein. He also gives an example of a Gorenstein ring which is not a complete intersection. He then proves that there is a unique element of R/M annihilated by the maximal ideal and that in the homogeneous case, if this element has degree d, there is a duality between elements of R/M of degree m and those of degree $d - m$. His proof assumes that the coefficient field is the field of real numbers, so that one can obtain dual elements which do not give zero when paired with the given element by requiring the corresponding coefficients in the dual basis to be equal to the original ones, so that the pairing will result in a sum of squares, which cannot be zero. However, this argument can be modified to give a proof which works in general. In addition to the algebraic generalizations we have been discussing, this theorem, together with the duality, has been used in the theory of residues in Algebraic Geometry, where it is often referred to as "Macaulay's Theorem", see Griffiths [5].

Generalizations of the concept of principal system have been as far-reaching and as basic to the subject as have those of depth and Cohen–Macaulay rings, and, as in that case, the subject took on new development with the advent of homological algebra. First, it has been generalized to arbitrary dimension in somewhat the same way that regular sequences have been, and is, as always, considered over arbitrary local rings. The main impetus to the development of this topic was the paper of Bass [2], where he generalized the condition to arbitrary dimension and showed that it was equivalent to the property that the ring has finite injective dimension. In the zero dimensional case, this says that the ring is its own injective hull, and the duality properties proven by Macaulay are then consequences of the injectivity of R/M as an R/M-module.

We mention one further development in this direction. We have already pointed out the generalization of the inverse system to the Matlis dual. This theory has also been generalized using more advanced methods of homological algebra to give a duality in the derived category of bounded complexes of modules with finitely generated homology, which is callled Grothendieck duality (see Hartshorne [6]). This theory has been extended to all subschemes of regular schemes, and Gorenstein schemes are characterized as those for which the dualizing complex is a locally free module, so locally generated by one element. While this is far from Macaulay's principal systems, it can be seen to be a direct analogue of Macaulay's condition that the inverse system be generated by one element.

Multiplicities.

While this subject is not treated extensively in this book, it is dealt with briefly and has become important later. The multiplicity of an ideal M such that R/M has finite length is the length of the localization of R/M at the maximal ideal $(x_1, \ldots, x_n)$. Macaulay defines the multiplicity using the inverse system, because in this way it can be defined as the dimension of a susbspace of the dual, avoiding the necessity of localizing the quotient module at the origin. The main theorem is that it the multiplicity of R/M agrees with that given by the resultant in the case in which M is a complete intersection. In addition, he proves some results on multiplicities for ideals which are residual with respect to a Gorenstein ideal (§§73–76). He states that the multiplicity has "no geometric significance" for ideals which are not complete intersections; however, this concept has since been generalized in several ways both in algebraic and geometric settings and its geometric significance for more general ideals is now well-established (see Samuel [20] and Serre [21]).

Perfect ideals.

After the discussion of primary modules, Macaulay takes up the subject of ideals of height less than the dimension of the ring. The main tool is the dialytic array mentioned above, and the main concept is that of "perfect module". He also considered "mutually residual modules".

While the computations in this section are quite complicated, we describe a part of the construction briefly. In considering a height r ideal M, usually assumed to be unmixed, Macaulay looks at it as an ideal in variables $x_1, \ldots x_r$ with coefficients in $k(x_{r+1}, \ldots, x_n)$, thus effectively localizing to reduce to the case where the quotient has finite length. This ideal is denoted $M^{(r)}$. Since the quotient now has finite length, there are a finite number of monomials such that every monomial can be written as a linear combination of these modulo $M^{(r)}$. However, he then examines the dyalitic array in detail and examines which elements of $k[x_{r+1}, \ldots, x_n]$ must be inverted to solve for the remaining monomials in terms of the basis of the quotient. The element which must be inverted to solve the equations is denoted R (§79). In particular, he describes the inverse system of $M^{(r)}$ quite explicitly. If nothing has to be inverted, so that $R = 1$, he calls the ideal perfect.

We recall that a module T is perfect in the modern sense if it has finite projective dimension and the projective dimension is equal to the length of a maximal regular sequence in its annihilator. For regular rings, using the connection between projective dimension and depth and the fact that every module has finite projective dimension, this condition is equivalent to the condition that T be Cohen–Macaulay. Thus for an ideal M in a polynomial ring, R/M is perfect if and only if R/M is Cohen–Macaulay. For general ideals, Macaulay's condition is stronger than the modern notion; for example he gives an example of an ideal generated by a regular sequence which is not perfect in his sense. However, for homogeneous ideals the two notions are equivalent. It is quite remarkable that he comes to the same idea by such a different route, and it is an interesting exercise to trace the connection between zero divisors in a system of parameters as in the modern definition and "extra rows" in the dialytic array as in Macaulay's (§77).

Macaulay also states the theorem that, if M is a homogeneous ideal and the quotient R/M is divided by a linear system of parameters, the multiplicity of the result gives the length of $R/M^{(r)}$ if and only if the ideal is perfect, and otherwise it is larger (§89). He does not prove this theorem here, as he evidently considered it to be obvious from his construction (a proof of a modern version can be found in Matsumura [15], Theorem 17.11). The idea here is that the length of the result of dividing by a system of parameters is the correct result if and only if the ring is Cohen–Macaulay. Much has been written on the problem of defining the multiplicity in such a way as to give the correct result in general since then. There have been several solutions to this question, of which we mention two. One approach is to define the multiplicity as a limit over powers of the ideal generated by the parameters as in Samuel [20]. A different method is to use the Euler characteristic of a Koszul complex on the parameters as in Serre [21]. Both methods give the same result, and both give Macaulay's answer in the case where R/M is Cohen–Macaulay.

We remark finally that near the end of the chapter there are two sections (§§86–87) on ideals which are mutually residual with respect to a complete intersection ideal. This is another subject which he treats rather briefly but which has had a great development in recent years. It is now called either liaison or linkage, and its present development started with the paper on liaison by Peskine and Szpiro [18]. Since then many other papers have been written on properties of linked ideals; for an account of more recent work on the subject we refer to the paper of Huneke and Ulrich [9].

Summary of Macaulay's Terminology.

In this section we summarize the main differences between the terminology used in Macaulay's book and that which is used today. We have not attempted to include those terms which do not have modern equivalents, such as "dialytic array" which are defined in the book.

Some of these terms have already been discussed in earlier sections. In particular, we have mentioned his use of the word "module" where today we would say "ideal". In the following list M and N denote ideals in a polynomial ring. We also recall that he used the term "spread" of an ideal to denote its set of solutions in affine space, and that he used the irreducible spread in place of the prime ideal in many cases. While these are not strictly speaking the same, we have used their equivalence to identify them in this list. Finally, we say "Cohen–Macaulay ideal" I to denote one for which $k[x_i]/I$ has this property; this is often used today. Also, in some cases the modern term has a more general meaning, and they are equivalent only in the cases considered by Macaulay.

Term used by Macaulay	*Modern term*
module	ideal
G.C.M of $M_1, \ldots, M_k$	$M_1 + \ldots + M_k$
L.C.M. of $M_1, \ldots, M_k$	$M_1 \cap \ldots \cap M_k$
M/N	$M : N$
M contains N	M is contained in N
dimensions	dimension
rank of M	height of M
simple module M	ideal M such that R/M is Artinian
H-module	homogeneous ideal
Noetherian module	ideal all of whose associated prime ideals are contained in $(x_1, \ldots x_n)$
relevant spread	associated prime ideal
spread of M	set of solutions of polynomials in Min k^n
module of the principal class	complete intersection ideal
principal system	Gorenstein ideal
perfect H-module	homogeneous ideal such that R/M is perfect
inverse system	Matlis dual
A-derivate of E (for E in the inverse system)	product of A and E

References

1. M. Auslander and D. Buchsbaum, *Homological dimension in local rings*, Trans. Amer. Math. Soc. **85** (1957), 390–405.
2. H. Bass, *On the ubiquity of Gorenstein rings*, Math. Zeit. **82** (1963), 8–28.
3. N. Bourbaki, *Algèbre Commutative*, Chapitres 5 à 7, Masson, 1985.
4. I. S. Cohen, *On the structure and ideal theory of complete local rings*, Trans. Amer. Math. Soc. **59**, (1946), 54–106.
5. P. Griffiths, *Complex Analysis and Algebraic Geometry*, Bull Amer. Math. Soc. new series 1 (1979), 595–626.
6. R. Hartshorne, *Residues and Duality*, Lecture Notes in Mathematics **20**, Springer 1965.
7. D. Hilbert, *Über die Theorie der algebraischen formen*, Math. Ann. **36** (1890) 471–534
8. M. Hochster, *Topics in the Homological Theory of modules over Commutative Rings*, Regional Conference series **24**, Amer. Math. Society, 1975
9. C. Huneke and B. Ulrich, *The structure of linkage*, Ann. of Math. **126** (1987), 277–334.
10. I. Kaplansky, *Commutative Rings*, First Jeffrey-Williams lecture, Canadian Mathematical Congress (1968).
11. W. Krull, *Idealtheorie*, Ergebnisse der Mathematik und ihrer Grenzgebiete, Springer-Verlag, Berlin 1935.
12. E. Lasker, *Zur Theorie der Moduln und Idealen*, Math. Ann. **60** (1905), 20–116.
13. F. S. Macaulay, *Some properties of enumeration in the theory of modular systems*, Proc. London Math. Soc. **26** (1927) 531–555
14. E. Matlis, *Injective modules over Noetherian Rings*, Pacific J. Math. **8** (1958), 511–528.
15. H. Matsumura, *Commutative Ring Theory*, Translated by M. Reid, Cambridge University Press, 1986.
16. E. Noether, *Idealtheorie in Ringbereichen*, Math. Ann **83** (1921), 24–66.
17. C. Peskine and L. Szpiro, *Dimension Projective finie et cohomologie locale*, Publ. Math. IHES **42** (1973), 47–119.
18. C. Peskine and L. Szpiro, *Liaison des varié tés algébriques*, Invent. Math. **26** (1974), 271–302.
19. P. Roberts, *Intersection theory and the homological conjectures in Commutative Algebra*, Proceedings of the International Congress of Mathematicians Kyoto 1990, 121–132, 1991.
20. P. Samuel, *La notion de multiplicité en Algèbre et en Géométrie Algébrique*, J. Math. Pures et Appliquèes, **30**, (1951) 159–274.
21. J.-P. Serre, *Algèbre locale-Multiplicités*, LNM **11**, Springer 1965

Paul Roberts
Salt Lake City

THE ALGEBRAIC THEORY OF MODULAR SYSTEMS

Introduction

Definition. A modular system is an infinite aggregate of polynomials, or whole functions* of n variables $x_1, x_2, ..., x_n$, defined by the property that if F, F_1, F_2 belong to the system $F_1 + F_2$ and AF also belong to the system, where A is any polynomial in $x_1, x_2, ..., x_n$.

Hence if F_1, F_2, ..., F_k belong to a modular system so also does $A_1 F_1 + A_2 F_2 + ... + A_k F_k$, where A_1, A_2, ..., A_k are arbitrary polynomials.

Besides the algebraic or relative theory of modular systems there is a still more difficult and varied absolute theory. We shall only consider the latter theory in so far as it is necessary for the former.

In the algebraic theory polynomials such as F and aF, where a is a quantity not involving the variables, are not regarded as different polynomials, and any polynomial of degree zero is equivalent to 1. No restriction is placed on the coefficients of F_1, F_2, ..., F_k except in so far as they may involve arbitrary parameters $u_1, u_2, ...,$ in which case they are restricted to being rational functions of such parameters. The same restriction applies to the coefficients of the arbitrary polynomials $A_1, A_2, ..., A_k$ above.

In the absolute theory the coefficients of F_1, F_2, ..., A_1, A_2, ... are restricted to a domain of integrity, generally ordinary integers or whole functions of parameters $u_1, u_2, ...$ with integral coefficients; and a polynomial of degree zero other than 1 or a unit is not equivalent to 1.

---

* We use the term *whole function* throughout the text (but not in the Note at the end) as equivalent to *polynomial* and as meaning a *whole rational function*.

Definitions. A modular system will be called a *module* (of poly-nomials).

Any polynomial F belonging to a module M is called a *member* (or element) of M.

According as we wish to denote that F is a member of M in the relative or absolute sense we shall write $F = 0 \bmod M$, or $F \equiv 0 \bmod M$. The notation $F \equiv 0 \bmod M$ only comes into use in the sequel in connection with the Resultant.

A *basis* of a module M is any set of members $F_1, F_2, ..., F_k$ such that every member of M is of the form $X_1 F_1 + X_2 F_2 + ... + X_k F_k$, where $X_1, X_2, ..., X_k$ are polynomials.

Every module of polynomials has a basis consisting of a finite number of members (Hilbert's theorem, § 37).

The proof of this theorem is from first principles, and its truth will be assumed throughout.

The theory of modular systems is very incomplete and offers a wide field for research. The object of the algebraic theory is to dis-cover those general properties of a module which will afford a means of answering the question whether a given polynomial is a member of a given module or not. Such a question in its simpler aspect is of im-portance in Geometry and in its general aspect is of importance in Algebra. The theory resembles Geometry in including a great variety of detached and disconnected theorems. As a branch of Algebra it may be regarded as a generalized theory of the solution of equations in several unknowns, and assumes that any given algebraic equation in one unknown can be completely solved. In order that a polynomial F may be a member of a module M whose basis $(F_1, F_2, ..., F_k)$ is given it is evident that F must vanish for all finite solutions (whether finite or infinite in number) of the equations $F_1 = F_2 = ... = F_k = 0$. These conditions are *sufficient* if M resolves into what are called *prime modules* *; otherwise they are not sufficient, and F must satisfy further conditions, also connected with the solutions, which may be difficult to express concretely. The first step is to find all the solutions of the equations $F_1 = F_2 = ... = F_k = 0$; and this is completely accomplished in the theories of the resultant and resolvent.

---

* Cayley and Salmon constantly assume this. Salmon also discusses particular cases of a number of important and suggestive problems connected with modular systems (Sa).

I. THE RESULTANT

1. The Resultant is defined in the first instance with respect to n *homogeneous* polynomials F_1, F_2, ..., F_n in n variables, of degrees l_1, l_2, ..., l_n, each polynomial being complete in all its terms with literal coefficients, all different. The resultant of any n given homogeneous polynomials in n variables is the value which the resultant in the general case assumes for the given case. The resultant of n given non-homogeneous polynomials in $n-1$ variables is the resultant of the corresponding homogeneous polynomials of the same degrees obtained by introducing a variable x_0 of homogeneity.

Definitions. An *elementary member* of the module $(F_1, F_2, ..., F_n)$ is any member of the type ωF_i $(i = 1, 2, ..., n)$, where ω is any power product of x_1, x_2, ..., x_n. What is and what is not an elementary member depends on the basis chosen for the module.

The total number of elementary members of an assigned degree is evidently finite.

The diagram below represents the array of the coefficients of all elementary members of $(F_1, F_2, ..., F_n)$ of degree t, arranged under the power products $\omega_1^{(t)}$, $\omega_2^{(t)}$, ..., $\omega_\mu^{(t)}$ of degree $t \left(\mu = \dfrac{\lfloor t+n-1}{\lfloor t \ \lfloor n-1} \right)$:

$$
\begin{array}{c|cccc}
 & \omega_1^{(t)} & \omega_2^{(t)} & \cdots\cdots\cdots & \omega_\mu^{(t)} \\
\hline
\lambda_1 & a_1 & b_1 & \cdots\cdots & k_1 \\
\lambda_2 & a_2 & b_2 & \cdots\cdots & k_2 \\
 & \multicolumn{4}{c}{\cdots\cdots\cdots\cdots\cdots} \\
\lambda_\rho & a_\rho & b_\rho & \cdots\cdots & k_\rho
\end{array}
$$

Each row of the array, in association with $\omega_1^{(t)}$, $\omega_2^{(t)}$, ..., $\omega_\mu^{(t)}$, represents an elementary member of degree t; and the rows of the array corresponding to F_i all consist of the same elements (the coefficients of F_i and zeros) but in different columns.

Any member $F = X_1 F_1 + X_2 F_2 + ... + X_n F_n$ of degree t is evidently a linear combination $\lambda_1 \omega_1 F_1 + \lambda_2 \omega_2 F_1 + ... + \lambda_p \omega_p F_i + ... + \lambda_\rho \omega_\rho F_n$ of elementary members of degree t, and is represented by the above array when bordered by λ_1, λ_2, ..., λ_ρ on the left, where λ_1, λ_2, ..., λ_ρ are the coefficients of X_1, X_2, ..., X_n, some of which may be zeros.

This bordered array also shows in a convenient way the whole coefficients of the terms of F, viz. $\Sigma \lambda a$, $\Sigma \lambda b$, ..., $\Sigma \lambda k$.

These remarks and definitions are equally applicable to any module $(F_1, F_2, ..., F_k)$ of homogeneous or non-homogeneous polynomials ; but the following definition applies only to the particular module $(F_1, F_2, ..., F_n)$.

The *resultant* R of $F_1, F_2, ..., F_n$ is the H.C.F. of the determinants of the above array for degree $t = l + 1$, where $l = l_1 + l_2 + ... + l_n - n$. It will be shown (§ 7) that R is homogeneous and of degree $l_1 l_2 ... l_n / l_i$ in the coefficients of F_i ($i = 1, 2, ..., n$).

2. Resultant of two homogeneous polynomials in two variables.

Let
$$F_1 = a_1 x_1^{l_1} + b_1 x_1^{l_1 - 1} x_2 + ... + k_1 x_2^{l_1},$$
$$F_2 = k_2 x_1^{l_2} + ... + a_2 x_2^{l_2},$$
$$l = l_1 + l_2 - 2.$$

The array of the coefficients of all elementary members of (F_1, F_2) of degree $l + 1$, viz. $x_1^{l_2 - 1} F_1, x_1^{l_2 - 2} x_2 F_1, ... x_2^{l_2 - 1} F_1, x_1^{l_1 - 1} F_2, ..., x_2^{l_1 - 1} F_2$, has l_2 rows corresponding to F_1 and l_1 rows corresponding to F_2, and the same number $l_1 + l_2$ of rows in all as columns. The resultant R is therefore the determinant of this array. The array is

$$
\begin{array}{c|cccccc}
& \omega_1^{(l+1)} & \omega_2^{(l+1)} & \cdots\cdots & \omega_{l_1+1}^{(l+1)} & \cdots\cdots\cdots & \omega_{l+2}^{(l+1)} \\
\hline
\lambda_1 & a_1 & b_1 \cdots\cdots\cdots k_1 & & . & . & . & = x_1^{l_2 - 1} F_1 \\
\lambda_2 & . & a_1 & b_1 \cdots\cdots\cdots k_1 & . & . & = x_1^{l_2 - 2} x_2 F_1 \\
& & & & & & \\
\lambda_{l_2} & . & . & a_1 & b_1 \cdots\cdots\cdots k_1 & . & = x_2^{l_2 - 1} F_1 \\
\lambda_{l_2+1} & k_2 \cdots\cdots\cdots\cdots a_2 & . & . & . & = x_1^{l_1 - 1} F_2 \\
& . & k_2 \cdots\cdots\cdots\cdots a_2 & . & . & = x_1^{l_1 - 2} x_2 F_2 \\
& & & & & & \\
\lambda_{l+2} & . & . & k_2 \cdots\cdots\cdots\cdots a_2 & = x_2^{l_1 - 1} F_2 \\
\end{array}
$$

On the right are written the elementary members which the rows represent. Thus, neglecting the left hand border, we may regard the diagram as a set of $l + 2$ identical equations for
$$\omega_1^{(l+1)}, \quad \omega_2^{(l+1)}, \quad ..., \quad \omega_{l+2}^{(l+1)}.$$
Solving them we have
$$R\omega_i^{(l+1)} = A_{i1} F_1 + A_{i2} F_2 \quad (i = 1, 2, ..., l + 2),$$
where A_{i1}, A_{i2} are polynomials whose coefficients are whole functions of the coefficients of F_1, F_2. Hence
$$R\omega^{(l+1)} \equiv 0 \bmod (F_1, F_2),$$

where $\omega^{(l+1)}$ is any power product of x_1, x_2 of degree $l+1$. This expresses the first important property of R.

3. Irreducibility of R. The general expression for the resultant R is irreducible in the sense that it cannot be resolved into two factors each of which is a whole function of the coefficients of F_1, F_2. When this has been proved it follows that any whole function of the coefficients of F_1, F_2 which vanishes as a consequence of R vanishing must be divisible by R.

R has a term $a_1^{l_2} a_2^{l_1}$ obtained from the diagonal of the determinant, and this is the only term of R containing $a_2^{l_1}$. Also, when $a_1 = 0$, R has a term $(-1)^{l_2} k_2 b_1^{l_2} a_2^{l_1-1}$, and this is the only term of R containing $a_2^{l_1-1}$ when $a_1 = 0$. Hence, when R is expanded in powers of a_2 to two terms, we have

$$R = a_1^{l_2} a_2^{l_1} + b a_2^{l_1-1} + \dots,$$

where $\qquad b \equiv (-1)^{l_2} k_2 b_1^{l_2} \bmod a_1.$

Hence if R can be written as a product of two factors, we have

$$R = (a_1^{p_1} a_2^{p_2} + \dots)(a_1^{q_1} a_2^{q_2} + \dots),$$

where $p_1 + q_1 = l_2$ and $p_2 + q_2 = l_1$, and either p_1 or q_1 is zero; for otherwise the coefficient b of $a_2^{l_1-1}$ would be zero or divisible by a_1, which is not the case. Hence one of the factors of R is independent of the coefficients of F_1, since both factors must be homogeneous in the coefficients of F_1. Similarly one of the factors must be independent of the coefficients of F_2, i.e.

$$R = (a_1^{l_2} + \dots)(a_2^{l_1} + \dots) = a_1^{l_2} a_2^{l_1},$$

since the whole coefficient of $a_1^{l_2}$ in R is $a_2^{l_1}$, and of $a_2^{l_1}$ is $a_1^{l_2}$. This is not true; hence R is irreducible.

4. *The necessary and sufficient condition that the equations $F_1 = F_2 = 0$ may have a proper solution (i.e. a solution other than $x_1 = x_2 = 0$) is the vanishing of R.*

This is the fundamental property of the resultant. If the equations $F_1 = F_2 = 0$ have a solution other than $x_1 = x_2 = 0$ it follows from

$$Rx_1^{l+1} \equiv 0 \bmod (F_1, F_2), \qquad Rx_2^{l+1} \equiv 0 \bmod (F_1, F_2),$$

that $R = 0$, by giving to x_1, x_2 the values (not both zero) which satisfy the equations $F_1 = F_2 = 0$.

Conversely if $R = 0$ we can choose $\lambda_1, \lambda_2, \dots, \lambda_{l+2}$ so that the sum of their products with the elements in each column of the determinant R vanishes. Multiplying each sum by the power product corresponding to its column, and adding by rows, we have

$$(\lambda_1 x_1^{l_2-1} + \lambda_2 x_1^{l_2-2} x_2 + \dots + \lambda_{l_2} x_2^{l_2-1}) F_1$$
$$+ (\lambda_{l_2+1} x_1^{l_1-1} + \dots + \lambda_{l+2} x_2^{l_1-1}) F_2 = 0,$$

where $\lambda_1, \lambda_2, \dots \lambda_{l+2}$ do not all vanish. Hence, since $\lambda_1 x_1^{l_2-1} + \dots$ is of less degree than F_2, F_1 must have a factor in common with F_2, and the equations $F_1 = F_2 = 0$ have a proper solution.

In the following article another proof is given which can be extended more easily to any number of variables.

5. When $R \neq 0$ there are $l + 2$ linearly independent members of (F_1, F_2) of degree $l + 1$, and l of degree l. When $R = 0$ there are only $l + 1$ linearly independent members of degree $l + 1$ and still l of degree l, i.e. in each case 1 less than the number of terms in a polynomial of degree $l + 1$ and l respectively. Hence there will be one and only one identical linear relation between the coefficients of the general member of (F_1, F_2) whether of degree $l + 1$ or l. Let this identical relation for degree $l + 1$ be

$$c_{l+1,0} z_{l+1,0} + c_{l,1} z_{l,1} + \dots + c_{0,l+1} z_{0,l+1} = 0,$$

where $z_{i,j}$ denotes the coefficient of $x_1^i x_2^j$ in the general member of (F_1, F_2) of degree $i + j$, and the $c_{i,j}$ are constants. Then, if F is the general member

$$z_{l,0} x_1^l + z_{l-1,1} x_1^{l-1} x_2 + \dots + z_{0,l} x_2^l$$

of (F_1, F_2) of degree l, $x_1 F$ is a member of degree $l + 1$ whose coefficients must satisfy the relation above. Hence

$$c_{l+1,0} z_{l,0} + c_{l,1} z_{l-1,1} + \dots + c_{1,l} z_{0,l} = 0.$$

Similarly $\quad c_{l,1} z_{l,0} + c_{l-1,2} z_{l-1,1} + \dots + c_{0,l+1} z_{0,l} = 0,$

since $x_2 F$ is a member of (F_1, F_2) of degree $l + 1$. These two relations must be equivalent to one only, since only one identical relation exists for degree l. Hence we have

$$\frac{c_{l+1,0}}{c_{l,1}} = \frac{c_{l,1}}{c_{l-1,2}} = \dots = \frac{c_{1,l}}{c_{0,l+1}} = \frac{a_1}{a_2} \text{ (say)},$$

i.e. $c_{l+1,0}, c_{l,1}, \dots, c_{0,l+1}$ are proportional to $a_1^{l+1}, a_1^l a_2, \dots, a_2^{l+1}$. Hence the original identical relation may be written

$$z_{l+1,0} a_1^{l+1} + z_{l,1} a_1^l a_2 + \dots + z_{0,l+1} a_2^{l+1} = 0,$$

showing that the general member $z_{l+1,0} x_1^{l+1} + \dots$ of (F_1, F_2) of degree $l + 1$ vanishes when $x_1 = a_1$, $x_2 = a_2$, and that the equations $F_1 = F_2 = 0$ have the proper solution (a_1, a_2). The theorem being thus proved true in general is assumed to be true in particular.

6. Resultant of n homogeneous polynomials in n variables.

The general theory of the resultant to be now given is exactly parallel to that already given for two variables, although it involves points of much greater difficulty as might be expected. Another method of exposition depending on a different definition of the resultant is given in (K, p. 260 ff.).

Let $F_1, F_2, ..., F_n$ be n homogeneous polynomials of degrees $l_1, l_2, ..., l_n$ of which all the coefficients are different letters. In particular, let $a_1, a_2, ..., a_n$ be the coefficients of $x_1^{l_1}, x_2^{l_2}, ..., x_n^{l_n}$ in $F_1, F_2, ..., F_n$ respectively, and $c_1, c_2, ..., c_n$ the constant terms of $F_1, F_2, ..., F_n$ when x_n is put equal to 1, so that $c_n = a_n$. Let

$$l = l_1 + l_2 + ... + l_n - n, \quad L = l_1 l_2 ... l_n, \quad L_1 = L/l_1, \quad L_2 = L/l_2, \quad ... L_n = L/l_n.$$

The resultant R of $F_1, F_2, ..., F_n$ has already been defined (§ 1) as the H.C.F. of the determinants of the array of the coefficients of all elementary members of $(F_1, F_2, ..., F_n)$ of degree $l+1$.

We shall first consider a particular determinant D of the array, viz. that representing (§ 1) the polynomial

$$X^{(0)} F_1 + X^{(1)} F_2 + ... + X^{(n-1)} F_n \text{ of degree } l+1,$$

where $X^{(i)}$ denotes a polynomial in which all terms divisible by $x_1^{l_1}$ or $x_2^{l_2} ...$ or $x_i^{l_i}$ are absent, which may be expressed by saying that $X^{(i)}$ is *reduced* in $x_1, x_2, ..., x_i$. The polynomial

$$X^{(0)} F_1 + X^{(1)} F_2 + ... + X^{(n-1)} F_n$$

is represented by the bordered array

$$
\begin{array}{c|cccc|l}
 & \omega_1^{(l+1)} & \omega_2^{(l+1)} & \cdots\cdots\cdots & \omega_\mu^{(l+1)} & \\
\hline
\lambda_1 & a_1 & b_1 & \cdots\cdots\cdots & k_1 & = \omega_1 F_1 \\
\lambda_2 & a_2 & b_2 & \cdots\cdots\cdots & k_2 & = \omega_2 F_1 \\
\multicolumn{5}{c}{\cdots\cdots\cdots\cdots\cdots\cdots} \\
\lambda_\mu & a_\mu & b_\mu & \cdots\cdots\cdots & k_\mu & = \omega_\mu F_n \\
\end{array}
$$

where $\omega_1^{(l+1)}, \omega_2^{(l+1)}, ..., \omega_\mu^{(l+1)}$ are all the power products of $x_1, x_2, ..., x_n$ of degree $l+1$, and $\lambda_1, \lambda_2, ..., \lambda_\mu$ are the coefficients of $X^{(0)}, X^{(1)}, ..., X^{(n-1)}$. That this array has the same number μ of rows as columns is seen from the fact that *one and only one of the elements* $a_1, a_2, ..., a_n$* (*the coefficients of* $x_1^{l_1}, x_2^{l_2}, ..., x_n^{l_n}$ *in* $F_1, F_2, ..., F_n$) *occurs in each row and each column*. This is evident as regards the rows. To prove

---

* These are not the same as the $a_1, a_2, ..., a_n$ in the first column of the array. The latter should be represented by some other symbols.

that the same is true of the columns, we notice firstly that there is no power product $\omega^{(l+1)}$ of degree $l+1$ reduced in all the variables, for the highest power product of this kind is $x_1^{l_1-1} x_2^{l_2-1} \ldots x_n^{l_n-1}$ which is of degree $l < l+1$; and secondly, if we put every coefficient of $F_1, F_2, \ldots, F_n$, except only $a_1, a_2, \ldots, a_n$, equal to zero, the diagram will represent the polynomial

$$X^{(0)} a_1 x_1^{l_1} + X^{(1)} a_2 x_2^{l_2} + \ldots + X^{(n-1)} a_n x_n^{l_n},$$

in which each power product $\omega^{(l+1)}$ occurs once and once only, so that one and only one element $a_1, a_2, \ldots, a_n$ occurs in each column of D.

Thus D when expanded has a term $\pm a_1^{\mu_1} a_2^{\mu_2} \ldots a_n^{\mu_n}$, where μ_i is the number of terms in $X^{(i-1)}$, and by saying that the coefficient of this term in D is to be $+1$ we remove any ambiguity as to the sign of D. Also it is to be noted that *D vanishes when $c_1, c_2, \ldots, c_n$ all vanish,* for the column of D corresponding to x_n^{l+1} contains no elements other than $c_1, c_2, \ldots, c_n$ and zeros.

Regarding the diagram as giving μ identical equations for

$$\omega_1^{(l+1)}, \omega_2^{(l+1)}, \ldots, \omega_\mu^{(l+1)},$$

and solving, we have

$$D\omega^{(l+1)} \equiv 0 \bmod (F_1, F_2, \ldots, F_n),$$

where $\omega^{(l+1)}$ is any power product of $x_1, x_2, \ldots, x_n$ of degree $l+1$. It can be proved that the factors of D other than R can be divided out of this congruence equation, so that

$$R\omega^{(l+1)} \equiv 0 \bmod (F_1, F_2, \ldots, F_n);$$

but this will not be assumed in what follows*.

7. The number of rows in D corresponding to F_n is the number of terms in $X^{(n-1)}$. But $X^{(n-1)}$ is of degree $l+1-l_n$ or

$$(l_1 - 1) + (l_2 - 1) + \ldots + (l_{n-1} - 1),$$

and its terms consist of all the power products in

$$(1 + x_1 + \ldots + x_1^{l_1-1}) \ldots (1 + x_{n-1} + \ldots + x_{n-1}^{l_{n-1}-1})$$

each multiplied by a power of x_n; hence the number of the terms is $l_1 l_2 \ldots l_{n-1} = L_n$. Thus D is homogeneous and of degree L_n in the

---

* No proof of this has been published so far as I know. It can be proved that if A is any whole function of the coefficients of $F_1, F_2, \ldots, F_n$ not divisible by R, and $AF \equiv 0 \bmod (F_1, F_2, \ldots, F_n)$, then $F \equiv 0 \bmod (F_1, F_2, \ldots, F_n)$. Hence from $D\omega^{(l+1)} \equiv 0 \bmod (F_1, F_2, \ldots, F_n)$ we have $R\omega^{(l+1)} \equiv 0 \bmod (F_1, F_2, \ldots, F_n)$. The condition that A is not divisible by R is not needed if F is of degree $\leqslant l$.

coefficients of F_n, and homogeneous and of degree $> L_i$ in the coefficients of F_i $(i = 1, 2, ..., n-1)$. It follows that R, which is a factor of D, is at most of degree L_n in the coefficients of F_n. We shall prove that R is of this degree, and consequently of degree L_i in the coefficients of F_i.

Let D' be any other non-vanishing determinant of the array, viz.

$$\begin{array}{c|cccc} & \omega_1^{(l+1)} & \omega_2^{(l+1)} & \dots\dots\dots & \omega_\mu^{(l+1)} \\ \hline a_1 & a_1' & b_1' & \dots\dots\dots & k_1' \\ a_2 & a_2' & b_2' & \dots\dots\dots & k_2' \\ & \dots & \dots\dots\dots\dots & & \\ a_\mu & a_\mu' & b_\mu' & \dots\dots\dots & k_\mu' \end{array}$$

This represents the polynomial $A_1 F_1 + A_2 F_2 + ... + A_n F_n$, in which $a_1, a_2, ..., a_n$ are the (arbitrarily chosen) coefficients of $A_1, A_2, ..., A_n$ which are not zeros. Choose $\lambda_1, \lambda_2, ..., \lambda_\mu$ in the previous diagram so that we have identically

$$X^{(0)} F_1 + X^{(1)} F_2 + ... + X^{(n-1)} F_n = A_1 F_1 + A_2 F_2 + ... + A_n F_n.$$

This gives, by equating coefficients of power products on both sides,

$$\Sigma\lambda a = \Sigma a a', \quad \Sigma\lambda b = \Sigma a b', \quad ..., \quad \Sigma\lambda k = \Sigma a k'$$

as equations for $\lambda_1, \lambda_2, ..., \lambda_\mu$; and they have a unique solution, since D does not vanish.

Let $\begin{pmatrix} \lambda \\ a \end{pmatrix}$ denote the determinant of the substitution corresponding to the solution of the above equations for $\lambda_1, \lambda_2, ..., \lambda_\mu$ as linear functions of $a_1, a_2, ..., a_\mu$. Then if we put

$$\Sigma\lambda a = \Sigma a a' = \lambda_1', \quad \Sigma\lambda b = \Sigma a b' = \lambda_2', \quad ..., \quad \Sigma\lambda k = \Sigma a k' = \lambda_\mu'$$

we have

$$\begin{pmatrix} \lambda' \\ \lambda \end{pmatrix} = D, \quad \begin{pmatrix} \lambda' \\ a \end{pmatrix} = D', \quad \text{and} \quad \begin{pmatrix} \lambda' \\ \lambda \end{pmatrix}\begin{pmatrix} \lambda \\ a \end{pmatrix} = \begin{pmatrix} \lambda' \\ a \end{pmatrix}, \quad \text{i.e.} \quad D\begin{pmatrix} \lambda \\ a \end{pmatrix} = D',$$

by the rule of successive substitutions, or the rule for multiplying determinants. Hence

$$\frac{D'}{D} = \begin{pmatrix} \lambda \\ a \end{pmatrix}.$$

Now we can find the solution for $\lambda_1, \lambda_2, ..., \lambda_\mu$, or the solution of

$$X^{(0)} F_1 + X^{(1)} F_2 + ... + X^{(n-1)} F_n = A_1 F_1 + A_2 F_2 + ... + A_n F_n,$$

in the following way. First solve the equation

$$Y^{(0)} F_1 + Y^{(1)} F_2 + ... + Y^{(n-2)} F_{n-1} + X^{(n-1)} = A_n$$

for the unknowns $Y^{(0)}, Y^{(1)}, ..., Y^{(n-2)}, X^{(n-1)}$. This equation has a unique solution, since the more particular equation

$$Y^{(0)} x_1^{l_1} + Y^{(1)} x_2^{l_2} + ... + Y^{(n-2)} x_{n-1}^{l_{n-1}} + X^{(n-1)} = A_n$$

has a unique solution (for any given polynomial A_n can be expressed in one and only one way in the form on the left) and shows that the number of the coefficients of $Y^{(0)}, Y^{(1)}, ..., Y^{(n-2)}, X^{(n-1)}$ is equal to the number of equations they have to satisfy.

Substituting the value thus found for $X^{(n-1)}$ in the equation

$$X^{(0)} F_1 + X^{(1)} F_2 + ... + X^{(n-1)} F_n = A_1 F_1 + A_2 F_2 + ... + A_n F_n,$$

it becomes

$$X^{(0)} F_1 + X^{(1)} F_2 + ... + X^{(n-2)} F_{n-1}$$
$$= (A_1 + Y^{(0)} F_n) F_1 + ... + (A_{n-1} + Y^{(n-2)} F_n) F_{n-1},$$

where $Y^{(0)}, Y^{(1)}, ..., Y^{(n-2)}$ have been found. Next solve the equation

$$Z^{(0)} F_1 + Z^{(1)} F_2 + ... + Z^{(n-3)} F_{n-2} + X^{(n-2)} = A_{n-1} + Y^{(n-2)} F_n,$$

which has a unique solution for $Z^{(0)}, Z^{(1)}, ..., Z^{(n-3)}, X^{(n-2)}$. We can proceed in this way till $X^{(0)}, X^{(1)}, ..., X^{(n-1)}$, i.e. $\lambda_1, \lambda_2, ..., \lambda_\mu$, have all been found.

In this method of solving the unknowns on the left are associated with $F_1, F_2, ..., F_{n-1}$ only and not with F_n. Hence $\binom{\lambda}{a}$ is a rational function of the coefficients of $F_1, F_2, ..., F_n$ whose denominator is independent of the coefficients of F_n, and the same is therefore true of $\dfrac{D'}{D} = \binom{\lambda}{a}$. Hence every determinant D' of the array has a factor in common with D which is of degree L_n in the coefficients of F_n. The resultant R, which is the H.C.F. of all the determinants D', is therefore of degree L_i in the coefficients of F_i $(i = 1, 2, ..., n)$.

If we put $D = AR$, A is called the *extraneous factor* of D. We have proved that A is independent of the coefficients of F_n; and it is proved at the end of § 8 that A depends only on the coefficients of $(F_1, F_2, ..., F_{n-1})_{x_n=0}$.

8. Properties of the Resultant. Since D has a term $a_1^{\mu_1} ... a_n^{\mu_n}$ (§ 6) R *has a term* $a_1^{L_1} a_2^{L_2} ... a_n^{L_n}$. This is called the *leading term* of R.

Since D *vanishes when* $c_1, c_2, ..., c_n$ *all vanish* (§ 6) *the same is true of* R; for $D = AR$ and A is independent of $c_1, c_2, ..., c_n$.

The extraneous factor A *of* D *is a minor of* D, *viz. the minor obtained by omitting all the columns of* D *corresponding to power* *products reduced in $n-1$ of the variables and the rows which contain the elements $a_1, a_2, ..., a_n$ in the omitted columns $(M_2, p.$ 14). Thus D/A, where A is this minor of D, is an explicit expression for R.*

Each coefficient a of $F_1, F_2, ..., F_n$ is said to have a certain numerical *weight*, equal to the index of the power of one particular variable (say x_n) in the term of which a is the coefficient. In the case of non-homogeneous polynomials the variable chosen is generally the variable x_0 of homogeneity. Also the weight of a^p is defined as p times the weight of a, and the weight of $a^p b^q c^r ...$ as the sum of the weights of $a^p, b^q, c^r, ...$. A whole function of the coefficients is said to be *isobaric* when all its terms are of the same weight.

The resultant is isobaric and of weight L. Assign to $x_1, x_2, ..., x_n$ the weights $0, 0, ..., 0, 1$. Then the coefficients of $F_1, F_2, ..., F_n$ have the same weights as the power products of which they are the coefficients. The ith row of the determinant D represents the polynomial $\omega_i F_j = a_i \omega_1^{(l+1)} + b_i \omega_2^{(l+1)} + ... + k_i \omega_\mu^{(l+1)}$. Thus the weights of $a_i, b_i, ..., k_i$ are less than the weights of $\omega_1^{(l+1)}, \omega_2^{(l+1)}, ..., \omega_\mu^{(l+1)}$ respectively by the same amount, viz. the weight of ω_i. Hence, on expanding D, the weight of any term is less than the sum of the weights of $\omega_1^{(l+1)}, \omega_2^{(l+1)}, ..., \omega_\mu^{(l+1)}$ by the sum of the weights of $\omega_1, \omega_2, ..., \omega_\mu$; i.e. D is isobaric. Again, if in D each letter a is changed to au^q, where q is the weight of a, D becomes Du^w, where w is the weight of D; and consequently if D be expressed as a product of whole factors each factor must be isobaric. Thus R is isobaric and its weight is that of its leading term $a_1^{L_1} a_2^{L_2} ... a_n^{L_n}$, which is $l_n L_n = L$. The weight of D is the weight of $a_1^{\mu_1} a_2^{\mu_2} ... a_n^{\mu_n}$, which is also L, since $\mu_n = L_n$.

The whole coefficient of $a_1^{L_1} a_2^{L_2} ... a_{n-1}^{L_{n-1}}$ in R is $a_n^{L_n}$. For the coefficient must be a whole function of the coefficients of F_n only of degree L_n and weight $l_n L_n$, and a_n is the only coefficient of F_n of weight l_n.

A more general result (§ 9) is that *the whole coefficient of $a_n^{L_n}$ in R is $R_n^{l_n}$ where R_n is the resultant of* $(F_1, F_2, ..., F_{n-1})_{x_n=0}$. Hence also *the whole coefficient of $a_r^{L_r} a_{r+1}^{L_{r+1}} ... a_n^{L_n}$ is $R_r^{l_r \, l_{r+1}...l_n}$ where R_r is the resultant of* $(F_1, F_2, ..., F_{r-1})_{x_r=...=x_n=0}$.

Since the weights of D and R are the same *the weight of the extraneous factor A of D is zero.* This, taken in conjunction with the fact that A is independent of the coefficients of F_n, shows that A is a *whole function of the coefficients of $(F_1, F_2, ..., F_{n-1})_{x_n=0}$ only.*

9. *The resultant of $F_1, F_2, \ldots, F_n$ is irreducible and invariant.* It has been proved that the resultant is irreducible when $n = 2$ (§ 3); and the proof can be extended to the general case by induction.

Let $R_n =$ the resultant of $(F_1, F_2, \ldots, F_{n-1})_{x_n=0}$;

$F_0 =$ the resultant of the homogeneous polynomials $F_1^{(0)}$, $F_2^{(0)}, \ldots, F_{n-1}^{(0)}$ in $x_1, x_2, \ldots, x_{n-2}, x_0$ obtained from F_1, $F_2, \ldots, F_{n-1}$ by changing x_{n-1}, x_n to $x_{n-1}x_0, x_nx_0$;

$F_n' = (F_n)_{x_1 = \ldots = x_{n-2}=0} = k_n x_{n-1}^{l_n} + \ldots + a_n x_n^{l_n}$;

$R' =$ the resultant of $F_1, F_2, \ldots, F_{n-1}, F_n'$;

$R_0 =$ the resultant of F_0, F_n', two polynomials in x_{n-1}, x_n ;

$L_1' l_1 = L_2' l_2 = \ldots = L'_{n-1} l_{n-1} = l_1 l_2 \ldots l_{n-1} = L_n.$

Finally let $a_1, a_2, \ldots, a_n, c_1, c_2, \ldots, c_{n-1}$ denote the same coefficients of $F_1, F_2, \ldots, F_n$ as in § 6. We assume R_n irreducible and have to prove that R is irreducible.

F_0 is of weight L_n in the coefficients of $F_1^{(0)}, F_2^{(0)}, \ldots, F_{n-1}^{(0)}$ and each coefficient is a homogeneous polynomial in x_{n-1}, x_n of degree equal to its weight in reference to x_0. Hence

$$F_0 = A x_{n-1}^{L_n} + B x_{n-1}^{L_n - 1} x_n + \ldots,$$

where $A, B, \ldots$ are whole functions of the coefficients of $F_1, F_2, \ldots, F_{n-1}$ of the same dimensions as R_n. When $x_n = 0$, F_0 becomes the resultant of $(F_1^{(0)}, F_2^{(0)}, \ldots, F_{n-1}^{(0)})_{x_n=0}$, viz. $R_n x_{n-1}^{L_n}$; hence $A = R_n$. Also the whole coefficient of $a_1^{L_1'} a_2^{L_2'} \ldots a_{n-2}^{L'_{n-2}}$ in F_0 is $a'^{L'_{n-1}}_{n-1}$, where a'_{n-1} is the coefficient of $x_0^{l_{n-1}}$ in $F_{n-1}^{(0)}$ (§ 8), viz.

$$a'_{n-1} = a_{n-1} x_{n-1}^{l_{n-1}} + b x_{n-1}^{l_{n-1} - 1} x_n + \ldots + c_{n-1} x_n^{l_{n-1}}.$$

Hence B has a term $L'_{n-1} a_1^{l_n'} \ldots a_{n-2}^{L'_{n-2}} a_{n-1}^{L'_{n-1}-1} b$, and cannot be divisible by R_n, since R_n does not involve b. Hence we find that

$$F_0 = R_n x_{n-1}^{L_n} + B x_{n-1}^{L_n - 1} x_n + \ldots$$

where B is neither zero nor divisible by R_n.

Now if R' vanishes one of the solutions of $F_n' = 0$ for $x_{n-1} : x_n$ will be the same as in one of the solutions of $F_1 = \ldots = F_{n-1} = 0$ (§ 10), and will therefore be a solution of $F_0 = 0$; i.e. $R' = 0$ requires $R_0 = 0$, and R_0 is divisible by each irreducible factor of R'. But (§ 3)

$$R_0 = R_n^{l_n} a_n^{L_n} + B' a_n^{L_n - 1} + \ldots, \text{ where } B' \equiv (-1)^{l_n} k_n B^{l_n} \bmod R_n,$$

so that B' is neither zero nor divisible by R_n. Hence, as in § 3, R_0 has an irreducible factor of the form $R_n^{l_n} a_n^p + \ldots$, and has no other factor involving the coefficients of F_1, F_2, ..., F_{n-1}. This must therefore be a factor of R'.

Again R' is what R becomes when all the coefficients of F_n other than those of F_n' are put equal to zero. Hence R has an irreducible factor of the form $R_n^{l_n} a_n^q + ...$, where $q \geqslant p$. The remaining factor of R is independent of the coefficients of $F_1, F_2, ..., F_{n-1}$, and therefore also of the coefficients of F_n when $n > 2$. Hence R is irreducible.

It easily follows that R is invariant for a homogeneous linear substitution whose determinant $\left(\dfrac{x}{x'}\right)$ does not vanish. Suppose that $R = 0$ and that this is the only relation existing between the coefficients of $F_1, F_2, ..., F_n$. Then not more than one relation can exist between the coefficients of $F_1', F_2', ..., F_n'$, the polynomials into which $F_1, F_2, ..., F_n$ transform. Since $R = 0$ there are less than μ linearly independent members of $(F_1, F_2, ..., F_n)$ of degree $l + 1$ and therefore less than μ linearly independent members of $(F_1', F_2', ..., F_n')$ of degree $l + 1$, and the only single relation between the coefficients of $F_1', F_2', ..., F_n'$ which will admit this is $R' = 0$. Hence $R = 0$ requires $R' = 0$, and R' is divisible by R. The remaining factor of R' is independent of the coefficients of $F_1, F_2, ..., F_n$, and can be shown to be $\left(\dfrac{x}{x'}\right)^L$. A proof that R is invariant without assuming it irreducible is given in (E, p. 17).

10. *The necessary and sufficient condition that the equations $F_1 = F_2 = ... = F_n = 0$ may have a proper solution is the vanishing of R.*

In the general case, when the coefficients are letters,
$$A R x_n^{l+1} \equiv 0 \mod (F_1, F_2, ..., F_n).$$
Put $x_n = 1$ and change* c_i to $c_i - F_i$ $(i = 1, 2, ..., n)$; then A does not change, being independent of $c_1, c_2, ..., c_n$ (§ 8); but R changes to $R - A_1 F_1 - A_2 F_2 - ... - A_n F_n$, and this must vanish; hence
$$R \equiv 0 \mod (F_1, F_2, ..., F_n)_{x_n=1}.$$
Hence R vanishes if the equations $F_1 = F_2 = ... = F_n = 0$ have a solution in which $x_n = 1$, i.e. if they have a proper solution.

To prove that $R = 0$ is a sufficient condition, we shall assume that $R = 0$ is the only relation existing between the coefficients of $F_1, F_2, ..., F_n$. There are then less than μ linearly independent members of $(F_1, F_2, ..., F_n)$ of degree $l + 1$. Hence the coefficients $z_{p_1, p_2, ..., p_n}$ of

---

* Called the Kronecker substitution.

the general member of degree $l+1$ must satisfy an identical linear relation

$$\Sigma c_{p_1, p_2, \ldots, p_n} z_{p_1, p_2, \ldots, p_n} = 0, \quad p_1 + p_2 + \ldots + p_n = l+1.$$

The coefficients of the general member of degree l also satisfy one *and only one* identical linear relation, whether R vanishes or not. To prove this it has to be shown that the number N of linearly independent members of $(F_1, F_2, \ldots, F_n)$ of degree l is 1 less than the number ρ of power products of degree l. If no relation exists between the coefficients of $F_1, F_2, \ldots, F_n$ the equation

$$X^{(0)}F_1 + X^{(1)}F_2 + \ldots + X^{(n-1)}F_n = A_1F_1 + A_2F_2 + \ldots + A_nF_n$$

can always be solved by the method of § 7, where $A_1, A_2, \ldots, A_n$ are arbitrary given polynomials. Hence N is not greater than the number of coefficients in $X^{(0)}, X^{(1)}, \ldots, X^{(n-1)}$, or in

$$X^{(0)}x_1^{l_1} + X^{(1)}x_2^{l_2} + \ldots + X^{(n-1)}x_n^{l_n},$$

viz. $\rho - 1$, since, when this expression is of degree l, every power product except $x_1^{l_1-1}x_2^{l_2-1}\ldots x_n^{l_n-1}$ occurs once and only once in it. Hence $N \leqslant \rho - 1$.

Any particularity in $F_1, F_2, \ldots, F_n$ can only affect the value of N by diminishing it. Hence for the remainder of the proof it will be sufficient to show that $N = \rho - 1$ in a particular example in which $R = 0$. Let

$$F_1 = (x_1 - x_2)x_1^{l_1-1}, \quad F_2 = (x_2 - x_3)x_2^{l_2-1}, \ldots, \quad F_n = (x_n - x_1)x_n^{l_n-1}.$$

Then $R = 0$ since the equations $F_1 = F_2 = \ldots = F_n = 0$ have the proper solution $x_1 = x_2 = \ldots = x_n = 1$. Let $x_1^{p_1}x_2^{p_2}\ldots x_n^{p_n}$ be any power product of degree l. If $p_1 \geqslant l_1$ change $x_1^{p_1}x_2^{p_2}$ to $x_1^{l_1-1}x_2^{q_2}$ where $p_1 + p_2 = l_1 - 1 + q_2$; this is equivalent to changing $x_1^{p_1}x_2^{p_2}\ldots x_n^{p_n}$ to $x_1^{p_1}x_2^{p_2}\ldots x_n^{p_n} + A_1F_1$. Again if $q_2 \geqslant l_2$ change $x_2^{q_2}x_3^{p_3}$ to $x_2^{l_2-1}x_3^{q_3}$; and if $q_2 < l_2$ proceed to the first $p_r \geqslant l_r$ and change $x_r^{p_r}x_{r+1}^{p_{r+1}}$ to $x_r^{l_r-1}x_{r+1}^{q_{r+1}}$. If we continue this process, going round the cycle $x_1, x_2, \ldots, x_n$ as many times as is necessary, the power product $x_1^{p_1}x_2^{p_2}\ldots x_n^{p_n}$ will eventually become changed to $x_1^{l_1-1}x_2^{l_2-1}\ldots x_n^{l_n-1}$. Hence these two power products are congruent mod $(F_1, F_2, \ldots, F_n)$, while neither of them is a member of $(F_1, F_2, \ldots, F_n)$, since they do not vanish when $x_1 = \ldots = x_n = 1$. Hence $N = \rho - 1$.

Let $F = \Sigma z_{q_1, q_2, \ldots, q_n} x_1^{q_1}x_2^{q_2}\ldots x_n^{q_n}$ be the general member of $(F_1, F_2, \ldots, F_n)$ of degree l; then x_iF is a member of degree $l+1$ in which the coefficient of $x_1^{p_1}x_2^{p_2}\ldots x_n^{p_n}$ is $z_{p_1, p_2, \ldots, p_i-1, \ldots, p_n}$. Hence

$$\Sigma c_{p_1, p_2, \ldots, p_n} z_{p_1, p_2, \ldots, p_i-1, \ldots, p_n} = 0 \quad (i = 1, 2, \ldots, n),$$

or $$\Sigma c_{q_1, q_2, \ldots, q_i+1, \ldots, q_n} z_{q_1, q_2, \ldots, q_n} = 0 \quad (i = 1, 2, \ldots, n).$$

These n equations in $z_{q_1, q_2, \dots, q_n}$ are therefore equivalent to one only; and the continued ratio $c_{q_1+1, q_2, \dots, q_n} : c_{q_1, q_2+1, \dots, q_n} : \dots : c_{q_1, q_2, \dots, q_n+1}$ is the same for all sets of values of $q_1, q_2, \dots, q_n$ whose sum is l. Equating to $a_1 : a_2 : \dots : a_n$, it follows that $c_{p_1, p_2, \dots, p_n}$ is proportional to $a_1^{p_1} a_2^{p_2} \dots a_n^{p_n}$ $(p_1 + p_2 + \dots + p_n = l + 1)$. Hence it follows that $(a_1, a_2, \dots, a_n)$ is a solution of the equations $F_1 = F_2 = \dots = F_n = 0$.

11. The Product Theorem. *If F_n is the product of two polynomials $F_n{}', F_n{}''$, the resultant R of $F_1, F_2, \dots, F_n$ is the product of the resultants R', R'' of $F_1, F_2, \dots, F_n{}'$ and $F_1, F_2, \dots, F_n{}''$.*

For in the general case R' and R'' are irreducible, and if either vanishes R vanishes. Hence R is divisible by $R'R''$. Also it can be easily verified that the leading terms of R and $R'R''$ are identical. Hence $R = R'R''$.

This result can easily be extended to the case in which any or all of $F_1, F_2, \dots, F_n$ resolve into two or more factors.

If $F_1, F_2, \dots, F_n$ are all members of the module $(F_1{}', F_2{}', \dots, F_n{}')$ the resultant R of $F_1, F_2, \dots, F_n$ is divisible by the resultant R' of $F_1{}', F_2{}', \dots, F_n{}'$. For if $R' = 0$ then $R = 0$.

12. Solution of Equations by means of the Resultant. The method of the resultant for solving equations can only be applied in what is called the principal case, that is, the case in which the number r of the equations is not greater than the number n of the unknowns, and the resultant F_0 of the equations with respect to $x_1, x_2, \dots, x_{r-1}$ (after a linear substitution of the unknowns) does not vanish identically. When F_0 vanishes identically the method of the resultant fails, but the equations can be solved by the method of the resolvent, due to Kronecker, as explained later. The method of the resolvent is also applicable to any number of equations whether greater or less than the number of unknowns.

Homogeneous Equations. Let the equations be $F_1 = F_2 = \dots = F_r = 0$ of degrees $l_1, l_2, \dots, l_r$, where $r \leqslant n$. We assume that their resultant F_0 with respect to $x_1, x_2, \dots, x_{r-1}$ does not vanish. We regard $x_1, x_2, \dots, x_r$ as the unknowns, the solutions being functions of $x_{r+1}, \dots, x_n$. But instead of solving for one of the unknowns $x_1, x_2, \dots, x_r$ we solve for a linear combination of them, viz. for $x = u_1 x_1 + u_2 x_2 + \dots + u_r x_r$,* where $u_1, u_2, \dots, u_r$ are undetermined quantities. Let F_u stand for $x - u_1 x_1 - u_2 x_2 - \dots - u_r x_r$. Then we regard $F_1 = F_2 = \dots = F_r = F_u = 0$

---

* Called the Liouville substitution.

as the given system of equations with $x_1, ..., x_r, x$ as unknowns, and their resultant $F_0^{(u)}$ with respect to $x_1, x_2, ..., x_r$ gives the equation $F_0^{(u)} = 0$ for x.

Definition. $F_0^{(u)}$ is called the *u-resultant* of $(F_1, F_2, ..., F_r)$.

Applying the reasoning of § 9 it is seen that F_0 is the resultant (with respect to $x_1, ..., x_{r-1}, x_0$) of $F_1, F_2, ..., F_r$ when $x_r, x_{r+1}, ..., x_n$ are changed to $x_r x_0, x_{r+1} x_0, ..., x_n x_0$, and is a homogeneous polynomial in $x_r, x_{r+1}, ..., x_n$ of degree $L = l_1 l_2 ... l_r$, viz.

$$F_0 = R_{r+1} x_r^L + ...,$$

where R_{r+1} is the resultant of $(F_1, F_2, ..., F_r)_{x_{r+1}=...=x_n=0}$, and does not vanish; for a homogeneous substitution beforehand between $x_r, x_{r+1}, ..., x_n$ only would be carried through to F_0.

Similarly $F_0^{(u)}$ is the resultant (with respect to $x_1, ..., x_r, x_0$) of $F_1, F_2, ..., F_r, F_u$ when $x, x_{r+1}, ..., x_n$ are changed to $x x_0, x_{r+1} x_0, ..., x_n x_0$, and is a homogeneous polynomial $R'_{r+1} x^L + ...$ in $x, x_{r+1}, ..., x_n$ where R'_{r+1} is the resultant of $(F_1, F_2, ..., F_r, F_u)_{x_{r+1}=...=x_n=0}$. It is easily seen* that $R'_{r+1} = R_{r+1}$. Hence

$$F_0^{(u)} = R_{r+1} x^L + ..., \text{ where } R_{r+1} \neq 0.$$

To each solution $x_r = x_{ri}$ of $F_0 = 0$ corresponds a solution $(x_{1i}, x_{2i}, ..., x_{ri})$ of the equations $F_1 = F_2 = ... = F_r = 0$ for $x_1, x_2, ..., x_r$ (§ 10). There are therefore L solutions altogether, and they are all finite, since $R_{r+1} \neq 0$.

Similarly to each of the L solutions $x = x_i$ of $F_0^{(u)} = 0$ there corresponds a solution $(x_{1i}, x_{2i}, ..., x_{ri}, x_i)$ of $F_1 = ... = F_r = F_u = 0$; and as regards $(x_{1i}, x_{2i}, ..., x_{ri})$ the L solutions must be the same as those obtained by solving $F_0 = 0$. Hence it follows that

$$x_i = u_1 x_{1i} + u_2 x_{2i} + ... + u_r x_{ri},$$

where $x_{1i}, x_{2i}, ..., x_{ri}$ are independent of $u_1, u_2, ..., u_r$. Hence

$$F_0^{(u)} = R_{r+1} \Pi (x - u_1 x_{1i} - ... - u_r x_{ri}) \quad (i = 1, 2, ..., L).$$

Thus $F_0^{(u)}$ is a product of L factors which are linear in $x, u_1, u_2, ..., u_r$, and the coefficients of $u_1, u_2, ..., u_r$ in each factor supply a solution of the equations $F_1 = F_2 = ... = F_r = 0$.

Also the number of solutions is either $L = l_1 l_2 ... l_r$ or infinite, the latter being the case when F_0 vanishes identically.

---

* By introducing a as coefficient of x in F_u it is seen that R'_{r+1} is divisible by a^L by considering weight with respect to x. Also the whole coefficient of a^L in R'_{r+1} is R_{r+1} (§ 8). Hence $R'_{r+1} = a^L R_{r+1} = R_{r+1}$.

If D_u is the determinant for $(F_1, F_2, ..., F_r, F_u)$, regarding $x_1, x_2, ...,$ x_r, x_0 as the variables, like the D of § 6, we have $D_u = AF_0^{(u)}$. The extraneous factor A depends only on the coefficients of $(F_1, F_2, ..., F_r)_{x_0=0}$, that is, of $(F_1, F_2, ..., F_r)_{x_{r+1}=...=x_n=0}$. *Hence A is a pure constant, independent of $x_{r+1}, ..., x_n$ and of $u_1, u_2, ..., u_r$, and we may take $D_u = 0$ as the equation for x.*

Definition. The number of times a linear factor $x - u_1 x_{1t} - ... - u_r x_{rt}$ is repeated in $F_0^{(u)}$ or D_u is called the *multiplicity* of the solution $(x_{1t}, x_{2t}, ..., x_{rt})$. This term has a definite geometrical interpretation; it is the number of solutions or points, in the general case distinct, which come into coincidence with a particular solution or point in the particular example considered.

In the case of n homogeneous equations in n unknowns such that $R \neq 0$, the complete solution consists of the non-proper solution $(0, 0, ..., 0)$ with multiplicity $L = l_1 l_2 ... l_n$,

Non-homogeneous Equations. In the case of non-homogeneous equations a linear substitution beforehand affects only $x_1, x_2, ..., x_n$ and not the variable x_0 of homogeneity. Hence it is possible for R_{r+1} to vanish identically, while F_0 and $F_0^{(u)}$ do not, no matter what the original substitution may be. In this case there is a diminution in the number of finite solutions for x, but not in the number of linear factors of $F_0^{(u)}$. To a factor $u_1 x_{1t} + u_2 x_{2t} + ... + u_r x_{rt}$ of $F_0^{(u)}$ not involving x corresponds what is called an infinite solution of $F_1 = F_2 = ... = F_r = 0$ in the ratio $x_{1t} : x_{2t} : ... : x_{rt}$. Infinite solutions are however non-existent in the theory of modular systems (§ 42). An extreme case is that in which $F_0^{(u)}$ does not vanish identically, but is independent of x, when all the L solutions are at infinity.

It may happen that a system of non-homogeneous equations has only a finite number of finite solutions while the resultant F_0 vanishes identically. In such a case the method of the resultant fails to give the solutions.

Example. The equations $x_1^2 = x_2 + x_1 x_3 = x_3 + x_1 x_2 = 0$ have the finite solution $x_1 = x_2 = x_3 = 0$; but the resultant vanishes identically because the corresponding homogeneous equations

$$x_1^2 = x_0 x_2 + x_1 x_3 = x_0 x_3 + x_1 x_2 = 0$$

are satisfied by $x_0 = x_1 = 0$, a system of two independent equations only.

II. THE RESOLVENT

13. We shall follow, with some material deviations, König's exposition of Kronecker's method of solving equations by means of the resolvent. The equations are in general supposed to be non-homogeneous; and homogeneous equations are regarded as a particular case. Thus a homogeneous equation in n variables represents a cone of $n - 1$ dimensions with its vertex at the origin. *Homogeneous coordinates are excluded.*

The problem is to find all the solutions of any given system of equations $F_1 = F_2 = \dots = F_k = 0$ in n unknowns $x_1, x_2, \dots, x_n$. The unknowns are supposed if necessary to have been subjected to a homogeneous linear substitution beforehand, the object being to make the equations and their solutions of a general character, and to prevent any inconvenient result happening (such as an equation or polynomial being irregular* in any of the variables) which could have been avoided by a linear substitution at the beginning. In theoretical reasoning *this preliminary homogeneous substitution is always to be understood*; but is seldom necessary in dealing with a particular example.

The solutions we shall seek are (i) those, if any, which exist for x_1 when $x_2, x_3, \dots, x_n$ have arbitrary values; (ii) those which exist for x_1, x_2, not included in (i), when $x_3, \dots, x_n$ have arbitrary values; (iii) those which exist for x_1, x_2, x_3, not included in (i) or (ii), when $x_4, \dots, x_n$ have arbitrary values; and so on. A set of solutions for $x_1, x_2, \dots, x_r$ when $x_{r+1}, \dots, x_n$ have arbitrary values is said to be of *rank r*, and the *spread* of the points whose coordinates are the solutions is of *rank r* and *dimensions n − r*. If there are solutions of rank r and no solutions of rank $< r$ the system of equations $F_1 = F_2 = \dots = F_k = 0$ and the module $(F_1, F_2, \dots, F_k)$ are both said to be of rank r.

14. The polynomials $F_1, F_2, \dots, F_k$, and also all their factors are regular in x_1. Hence their common factor D can be found by the ordinary process of finding the H.C.F. of $F_1, F_2, \dots, F_k$ treated as polynomials in a single variable x_1. If D does not involve the variables we take it to be 1. If it does involve the variables the solutions of $D = 0$ treated as an equation for x_1 give the first set of solutions of the equations $F_1 = F_2 = \dots = F_k = 0$ mentioned above.

---

* A polynomial of degree l is said to be regular or irregular in x_1 according as the term x_1^l is present in it or not.

In the algebraic theory of modules we regard any algebraic equation in one unknown, whether the coefficients involve parameters or not, as completely soluble, i.e. we regard any given non-linear polynomial in one variable as *reducible*. A polynomial in two or more variables is called reducible if it is the product of two polynomials both of which involve the variables. A polynomial which is not reducible is called (absolutely) *irreducible*. Any given polynomial is either irreducible or uniquely expressible as a product of irreducible factors, leaving factors of degree zero out of account. It is assumed that the irreducible factors of any given polynomial are known. Thus the polynomial D above may be supposed to be expressed in its irreducible factors in $x_1, x_2, \ldots, x_n$, and to each irreducible factor corresponds an irreducible or non-degenerate spread.

Put $F_i = D\phi_i$ $(i = 1, 2, \ldots, k)$. Then $\phi_1, \phi_2, \ldots, \phi_k$ have no common factor involving the variables, and the same is true of the two polynomials

$$\lambda_1\phi_1 + \lambda_2\phi_2 + \ldots + \lambda_k\phi_k \text{ and } \mu_1\phi_1 + \mu_2\phi_2 + \ldots + \mu_k\phi_k,$$

where the λ's and μ's are arbitrary quantities. Regarding them as two polynomials in a single variable x_1 we calculate their resultant, and arrange it in the form

$$\rho_1 F_1^{(1)} + \rho_2 F_2^{(1)} + \ldots + \rho_{k_1} F_{k_1}^{(1)},$$

where $\rho_1, \rho_2, \ldots, \rho_{k_1}$ are different power products of the λ's and μ's, and $F_1^{(1)}, F_2^{(1)}, \ldots, F_{k_1}^{(1)}$ are polynomials in $x_2, x_3, \ldots, x_n$ not involving the λ's and μ's. Each $F_i^{(1)}$ is regular in x_2; for any homogeneous linear substitution beforehand of $x_2, x_3, \ldots, x_n$ among themselves only would be carried through to the $F_i^{(1)}$.

Find the H.C.F. $D^{(1)}$ of $F_1^{(1)}, F_2^{(1)}, \ldots, F_{k_1}^{(1)}$ treated as polynomials in a single variable x_2, and put $F_i^{(1)} = D^{(1)} \phi_i^{(1)}$ $(i = 1, 2, \ldots, k_1)$. Then find the resultant of

$$\lambda_1\phi_1^{(1)} + \lambda_2\phi_2^{(1)} + \ldots + \lambda_{k_1}\phi_{k_1}^{(1)} \text{ and } \mu_1\phi_1^{(1)} + \mu_2\phi_2^{(1)} + \ldots + \mu_{k_1}\phi_{k_1}^{(1)}$$

and arrange it in the form

$$\rho_1 F_1^{(2)} + \rho_2 F_2^{(2)} + \ldots + \rho_{k_2} F_{k_2}^{(2)}$$

as before, where $F_1^{(2)}, F_2^{(2)}, \ldots, F_{k_2}^{(2)}$ are polynomials in $x_3, x_4, \ldots, x_n$, which may be assumed regular in x_3, and whose H.C.F. $D^{(2)}$ can be found. We thus get the following series in succession :

$$F_1, \quad F_2, \quad \ldots, \quad F_k, \quad \text{with H.C.F. } D,$$
$$\phi_1, \quad \phi_2, \quad \ldots, \quad \phi_k,$$
$$F_1^{(1)}, F_2^{(1)}, \ldots, F_{k_1}^{(1)}, \text{ with H.C.F. } D^{(1)},$$
$$\phi_1^{(1)}, \phi_2^{(1)}, \ldots, \phi_{k_1}^{(1)},$$
$$F_1^{(2)}, F_2^{(2)}, \ldots, F_{k_2}^{(2)}, \text{ with H.C.F. } D^{(2)},$$
$$\phi_1^{(2)}, \phi_2^{(2)}, \ldots, \phi_{k_2}^{(2)}, \text{ and so on.}$$

Now any solution of $F_1 = F_2 = \ldots = F_k = 0$ is a solution of $D = 0$ or of $\phi_1 = \phi_2 = \ldots = \phi_k = 0$. And any solution of $\phi_1 = \phi_2 = \ldots = \phi_k = 0$ is a solution of $F_1^{(1)} = F_2^{(1)} = \ldots = F_{k_1}^{(1)} = 0$, since $\Sigma \rho_i F_i^{(1)} \equiv 0 \bmod (\Sigma \lambda_i \phi_i, \Sigma \mu_i \phi_i)$, and therefore a solution of $D^{(1)} = 0$ or of $\phi_1^{(1)} = \phi_2^{(1)} = \ldots = \phi_{k_1}^{(1)} = 0$. Hence any solution of $F_1 = F_2 = \ldots = F_k = 0$ is a solution of $D = 0$ or of $D^{(1)} = 0$ or of $\phi_1^{(1)} = \phi_2^{(1)} = \ldots = \phi_{k_1}^{(1)} = 0$. Proceeding in a similar way we find that any solution of $F_1 = \ldots = F_k = 0$ is a solution of $D D^{(1)} \ldots D^{(n-1)} = 0$, since $\phi_1^{(n-1)}, \phi_2^{(n-1)}, \ldots, \phi_{k_{n-1}}^{(n-1)}$ are polynomials in a single variable x_n at most and have no common factor.

Conversely if $\xi_3, x_4, \ldots, x_n$ is any solution of $D^{(2)} = 0$ the resultant of $\Sigma \lambda_i \phi_i^{(1)}$ and $\Sigma \mu_i \phi_i^{(1)}$ with respect to x_2 vanishes when $x_3 = \xi_3$, and $\Sigma \lambda_i \phi_i^{(1)} = \Sigma \mu_i \phi_i^{(1)} = 0$ have a solution $x_2 = \xi_2$ when $x_3 = \xi_3$; i.e. the equations $\phi_1^{(1)} = \ldots = \phi_{k_1}^{(1)} = 0$, and therefore also the equations $F_1^{(1)} = \ldots = F_{k_1}^{(1)} = 0$, have a solution $\xi_2, \xi_3, x_4, \ldots, x_n$; and, by the same reasoning, the equations $F_1 = F_2 = \ldots = F_k = 0$ have a solution $\xi_1, \xi_2, \xi_3, x_4, \ldots, x_n$. Similarly to any solution of $D D^{(1)} \ldots D^{(n-1)} = 0$, say a solution $\xi_i, x_{i+1}, \ldots, x_n$ of $D^{(i-1)} = 0$, there corresponds a solution $\xi_1, \xi_2, \ldots, \xi_i, x_{i+1}, \ldots, x_n$ of the equations $F_1 = F_2 = \ldots = F_k = 0$. Hence from the solutions of the single equation $D D^{(1)} \ldots D^{(n-1)} = 0$ we can get all the solutions of the system $F_1 = F_2 = \ldots = F_k = 0$, since all the solutions of the latter satisfy the former.

Definitions. $D D^{(1)} \ldots D^{(n-1)}$ is called the *complete (total) resolvent* of the equations $F_1 = F_2 = \ldots = F_k = 0$ and of the module $(F_1, F_2, \ldots, F_k)$. $D^{(i-1)}$ is called the *complete partial resolvent of rank i*, and any whole factor of $D^{(i-1)}$ is called a partial resolvent of rank i.

15. *The complete resolvent is a member of the module* $(F_1, F_2, \ldots, F_k)$.

For $\Sigma \rho_i F_i^{(1)} \equiv 0 \bmod (\Sigma \lambda_i \phi_i, \Sigma \mu_i \phi_i) = A \Sigma \lambda_i \phi_i + B \Sigma \mu_i \phi_i$,

where A, B are whole functions of $x_1, x_2, \ldots, x_n, \lambda_1, \ldots, \lambda_k, \mu_1, \ldots, \mu_k$. Hence by equating coefficients of the power products ρ_i on both sides, we have

$$F_i^{(1)} \equiv 0 \bmod (\phi_1, \phi_2, \ldots, \phi_k),$$

and $$D F_i^{(1)} = 0 \bmod (F_1, F_2, \ldots, F_k),^{*}$$

or $$D D^{(1)} \phi_i^{(1)} = 0 \bmod (F_1, F_2, \ldots, F_k).$$

Similarly $D D^{(1)} \ldots D^{(n-1)} \phi_i^{(n-1)} = 0 \bmod (F_1, F_2, \ldots, F_k)$;

and since the $\phi_i^{(n-1)}$ include one variable only (or none at all) and have

---

* Not $D F_i^{(1)} \equiv 0 \bmod (F_1, F_2, \ldots, F_k)$ because any common factor of $F_1, F_2, \ldots, F_k$ not involving the variables is not included in D and is left out of account.

no common factor, we can choose polynomials a_i in the single variable so that $\Sigma a_i \phi_i^{(n-1)} = 1$.   Hence

$$D D^{(1)} \dots D^{(n-1)} = 0 \bmod (F_1, F_2, \dots, F_k).$$

If the equations $F_1 = F_2 = \dots = F_k = 0$ have no finite solution the complete resolvent is equal to 1; consequently 1 is a member of $(F_1, F_2, \dots, F_k)$, and every polynomial is a member.

16.   We have seen that to every solution $x_i = \xi_i$ of $D^{(i-1)} = 0$ there corresponds a solution $\xi_1, \xi_2, \dots, \xi_i, x_{i+1}, \dots, x_n$ of the equations $F_1 = F_2 = \dots = F_k = 0$.   It may happen that there is an earlier complete partial resolvent $D^{(j-1)}$ which vanishes when $x_j = \xi_j, \dots, x_i = \xi_i$.   In such a case the solution $\xi_1, \dots, \xi_i, x_{i+1}, \dots, x_n$ of $F_1 = \dots = F_k = 0$ corresponding to a solution of $D^{(i-1)} = 0$ is included in the solutions corresponding to $D^{(j-1)} = 0$, and may be neglected if we are seeking merely the complete solution of $F_1 = F_2 = \dots = F_k = 0$.   Such a solution is called an *imbedded* solution.   All solutions corresponding to an irreducible factor of $D^{(i-1)}$ will be imbedded if one of them is imbedded.

17.   Examples on the Resolvent.   Geometrically the resolvent enables us to resolve the whole spread represented by any given set of algebraic equations into definite irreducible spreads (§ 21).   It has been supposed that the complete resolvent also supplies a definite answer to certain other questions.   The following examples disprove this to some extent.

Example i.   Find the resolvent of n homogeneous equations $F_1 = F_2 = \dots = F_n = 0$ of the same degree l and having no proper solution.

Since there are no solutions of rank $< n$ the complete resolvent is $D^{(n-1)}$.   The first derived set of polynomials $F_1^{(1)}, F_2^{(1)}, \dots, F_{k_1}^{(1)}$ are homogeneous and of degree l^2, the 2nd set $F_1^{(2)}, F_2^{(2)}, \dots$ are homogeneous and of degree l^4, and the $(n-1)$th set $F_1^{(n-1)}, F_2^{(n-1)}, \dots$ are homogeneous and of degree $l^{2^{n-1}}$.   This last set involve only one variable x_n, and therefore have the common factor $x_n^{l^{2^{n-1}}}$, which is therefore the required complete resolvent.

We should arrive at a similar result if we changed x_i to $x_i + a_i \, (i = 1, 2, \dots, n)$ beforehand, thus making the polynomials non-homogeneous.   The complete resolvent would then be $(x_n + a_n)^{l^{2^{n-1}}}$.   The resultant would be $(x_n + a_n)^{l^n}$.   The difference in the two results is explained by the fact that the resultant is obtained by a process applying uniformly to all the variables, and the resolvent by a process applied to the variables in succession.

Example ii.   König (K, p. 219) defines a module or system of equations as being *simple* or *mixed* according as only one or more than one of the complete partial resolvents D, $D^{(1)}$, ..., $D^{(n-1)}$ differs from unity. Kronecker (Kr, p. 31) says that the system of equations $F_1 = F_2 = ... = F_k = 0$ is irreducible in this case; and the *Ency. des Sc. Math.* (W$_2$, p. 352) repeats König's definition. We give two examples to show that this definition is a valueless one.

If u, v, w are three linear functions of three or more variables, any polynomial which contains the spread of $u = v = 0$ is of the form $Au + Bv$; if it also contains the spread of $u = w = 0$, B must vanish when $u = w = 0$, hence B must be of the form $Cu + Dw$, and $Au + Bv$ of the form $A'u + B'vw$; if it also contains the spread of $v = w = 0$, A' must be of the form $C'v + D'w$, and $A'u + B'vw$ of the form $C'uv + D'uw + B'vw$. Hence a polynomial which contains all three spreads is a member of the module (vw, wu, uv), and also any member of the module contains the three spreads. This module, although composite, is not mixed in any proper sense of the word.

Besides having partial resolvents of rank 2 corresponding to the three spreads the module has a partial resolvent of rank 3 corresponding to its singular spread $u = v = w = 0$. This last partial resolvent does not correspond to any property of the module which is not included in the properties corresponding to its partial resolvents of rank 2; in other words the partial resolvent of rank 3 is purely redundant.

The resolvent $D^{(1)} D^{(2)}$ can be found as follows : Suppose

$$u = a_0 + a_1 x_1 + a_2 x_2 + ..., \quad v = b_0 + b_1 x_1 + b_2 x_2 + ..., \quad w = c_0 + c_1 x_1 + c_2 x_2 + ...$$

Then the resultant of $\lambda_1 vw + \lambda_2 wu + \lambda_3 uv$ and $\mu_1 vw + \mu_2 wu + \mu_3 uv$ with respect to x_1, apart from a constant factor, is

$$(c_1 v - b_1 w)(a_1 w - c_1 u)(b_1 u - a_1 v)$$

$$\times \left\{ \frac{c_1 v - b_1 w}{\lambda_2 \mu_3 - \lambda_3 \mu_2} + \frac{a_1 w - c_1 u}{\lambda_3 \mu_1 - \lambda_1 \mu_3} + \frac{b_1 u - a_1 v}{\lambda_1 \mu_2 - \lambda_2 \mu_1} \right\},$$

its four irreducible factors corresponding to the spreads

$$v = w = 0, \quad w = u = 0, \quad u = v = 0,$$

$$(\lambda_2 \mu_3 - \lambda_3 \mu_2) u = (\lambda_3 \mu_1 - \lambda_1 \mu_3) v = (\lambda_1 \mu_2 - \lambda_2 \mu_1) w.$$

Hence $D^{(1)} = (c_1 v - b_1 w)(a_1 w - c_1 u)(b_1 u - a_1 v)$;

and $\phi_1^{(1)} = (c_1 v - b_1 w), \quad \phi_2^{(1)} = (a_1 w - c_1 u), \quad \phi_3^{(1)} = (b_1 u - a_1 v),$

from which we obtain

$$D^{(2)} = (b_1 c_2 - b_2 c_1)\, u + (c_1 a_2 - c_2 a_1)\, v + (a_1 b_2 - a_2 b_1)\, w.$$

Example iii. Compare and find the resolvents of the two modules

$$M = (x_1^3,\ x_2^3,\ x_1^2 + x_2^2 + x_1 x_2 x_3),$$
$$M' = (x_1^3,\ x_1^2 x_2,\ x_1 x_2^2,\ x_2^3,\ x_1^2 + x_2^2 + x_1 x_2 x_3).$$

The resolvent of M' will be found by obtaining the resultant with respect to x_1 of the two equations

$$\lambda_1 x_1^3 + \lambda_2 x_1^2 x_2 + \lambda_3 x_1 x_2^2 + \lambda_4 x_2^3 + \lambda_5 (x_1^2 + x_2^2 + x_1 x_2 x_3) = 0,$$

and $$\mu_1 x_1^3 + \mu_2 x_1^2 x_2 + \mu_3 x_1 x_2^2 + \mu_4 x_2^3 + \mu_5 (x_1^2 + x_2^2 + x_1 x_2 x_3) = 0.$$

This resultant is the same as that of the first equation and

$$(\lambda_1 \mu_5)\, x_1^3 + (\lambda_2 \mu_5)\, x_1^2 x_2 + (\lambda_3 \mu_5)\, x_1 x_2^2 + (\lambda_4 \mu_5)\, x_2^3 = 0$$

except for a factor λ_5^3. The roots of the last equation are $a_1 x_2,\ a_2 x_2,\ a_3 x_2$. Hence the resultant, apart from a constant factor, is

$$\Pi \left\{ (\lambda_1 a^3 + \lambda_2 a^2 + \lambda_3 a + \lambda_4)\, x_2^3 + \lambda_5 (a^2 + 1 + a x_3)\, x_2^2 \right\},\quad (a = a_1,\ a_2,\ a_3)$$

or $$x_2^6\, \Pi \left\{ (\lambda_1 a^3 + \lambda_2 a^2 + \lambda_3 a + \lambda_4)\, x_2 + \lambda_5 (a^2 + 1 + a x_3) \right\}.$$

Hence the complete resolvent is x_2^6, since no values of $x_2,\ x_3$ independent of the λ's and μ's will make the remaining product of factors of the above resultant vanish.

The complete resolvent of M, worked in the same way, is also x_2^6; i.e. M and M' have the same complete resolvent, although they are not the same module. M, but not M', contains the two modules

$$M'' = (x_3 - 1,\ x_1^2 + x_1 x_2 + x_2^2,\ x_1^2 x_2 + x_1 x_2^2),$$
$$M''' = (x_3 + 1,\ x_1^2 - x_1 x_2 + x_2^2,\ x_1^2 x_2 - x_1 x_2^2),$$

i.e. every member of M is a member of M'' and of M'''. Thus

$$x_1^3 = x_1(x_1^2 + x_1 x_2 + x_2^2) - (x_1^2 x_2 + x_1 x_2^2),$$
$$x_2^3 = x_2(x_1^2 + x_1 x_2 + x_2^2) - (x_1^2 x_2 + x_1 x_2^2),$$
$$x_1^2 + x_2^2 + x_1 x_2 x_3 = (x_1^2 + x_1 x_2 + x_2^2) + x_1 x_2(x_3 - 1).$$

The module M is what is called the L.C.M. of $M',\ M'',\ M'''$. The two modules $M'',\ M'''$ have $x_1 = x_2 = x_3 - 1 = 0$ and $x_1 = x_2 = x_3 + 1 = 0$ for their spreads, which are imbedded in the spread $x_1 = x_2 = 0$ of the first component of M, viz. M'.

M is then properly speaking a mixed module although this is not indicated by its complete resolvent x_2^6. It has two imbedded spreads, the points $(0, 0, \pm 1)$. The complete resolvent should have the factors

$x_3 \pm 1$ to indicate these, but it has no such factors. The complete resolvent may indicate imbedded modules which do not exist as in Ex. ii, or it may give no indication of them when they do exist as in Ex. iii.

Example iv. It is stated in the *Encyk. der Math. Wiss.* (W_1, p. 305) and repeated in (W_2, p. 354) that if only one complete partial resolvent $D^{(r)}$ differs from 1, and $D^{(r)}$ has no repeated factor, the module is the product of the prime modules corresponding to the irreducible factors of $D^{(r)}$. The absurdity of this statement is shown by applying it to the module (u, vw), where u, v, w are the same as in Ex. ii. The complete resolvent is $D^{(1)} = (b_1 u - a_1 v)(c_1 u - a_1 w)$, and the product of the prime modules $(u, v), (u, w)$ corresponding to its two factors is $(u^2, uv, uw, vw) \neq (u, vw)$.

18. The u-resolvent. The solutions of $F_1 = F_2 = \ldots = F_k = 0$ are obtained in the most useful way by introducing a general unknown x standing for $u_1 x_1 + u_2 x_2 + \ldots + u_n x_n$, where $u_1, u_2, \ldots, u_n$ are undetermined coefficients. This is done by putting

$$x_1 = \frac{x - u_2 x_2 - \ldots - u_n x_n}{u_1}$$

in the system of equations $F_1 = F_2 = \ldots = F_k = 0$. We thus get a new system $f_1 = f_2 = \ldots = f_k = 0$ in $x, x_2, x_3, \ldots, x_n$, where

$$f_i = u_1{}^{l_i} F_i \left(\frac{x - u_2 x_2 - \ldots - u_n x_n}{u_1}, x_2, \ldots, x_n \right) \quad (i = 1, 2, \ldots, k),$$

the multiplier $u_1{}^{l_i}$ being introduced to make f_i integral in u_1. There is evidently a one-one correspondence between the solutions of the two systems, viz. to the solution $\xi_1, \xi_2, \ldots, \xi_n$ of $F_1 = F_2 = \ldots = F_k = 0$ there corresponds the solution $\xi, \xi_2, \ldots, \xi_n$ of $f_1 = f_2 = \ldots = f_k = 0$, and *vice versa*, where $\xi = u_1 \xi_1 + u_2 \xi_2 + \ldots + u_n \xi_n$.

Definition. The complete resolvent $D_u D_u{}^{(1)} \ldots D_u{}^{(n-1)} (= F_u)$ of $(f_1, f_2, \ldots, f_k)$ obtained by eliminating $x_2, x_3, \ldots, x_n$ in succession is called the *complete u-resolvent* of $(F_1, F_2, \ldots, F_k)$.

Since $F_u = 0 \mod (f_1, f_2, \ldots, f_k)$, by § 15, we have

$$(F_u)_{x = u_1 x_1 + \ldots + u_n x_n} = 0 \mod (F_1, F_2, \ldots, F_k).$$

F_u is a whole function of $x, x_2, \ldots, x_n, u_1, u_2, \ldots, u_n$ which resolves into linear factors when regarded as a function of x only. The linear factors of rank r, that is, the linear factors of $D_u{}^{(r-1)}$, are of the type

$$x - u_1 \xi_1 - \ldots - u_r \xi_r - u_{r+1} x_{r+1} - \ldots - u_n x_n$$

where ξ_1, ..., ξ_r, x_{r+1}, ..., x_n is a solution of $F_1 = F_2 = ... = F_k = 0$. For if $x - \xi$ is any linear factor of $D_u^{(r-1)}$ then ξ is a root of $D_u^{(r-1)} = 0$ to which corresponds a solution $\xi, \xi_2, ..., \xi_r, x_{r+1}, ..., x_n$ of $f_1 = f_2 = ... = f_k = 0$ (§ 14) and a solution $\xi_1, \xi_2, ..., \xi_r, x_{r+1}, .., x_n$ of $F_1 = F_2 = ... = F_k = 0$, where $\xi = u_1\xi_1 + ... + u_r\xi_r + u_{r+1}x_{r+1} + ... + u_n x_n$.

The linear factors of F_u expressed in the above form supply all the solutions of $f_1 = f_2 = ... = f_k = 0$, viz. $\xi, \xi_2, ..., \xi_r, x_{r+1}, ..., x_n$, and all the solutions of $F_1 = F_2 = ... = F_k = 0$, viz. $\xi_1, \xi_2, ..., \xi_r, x_{r+1}, ..., x_n$, of the several ranks $r = 1, 2, ..., n$; but it is only when $\xi_1, \xi_2, ..., \xi_r$ are independent of $u_1, u_2, ..., u_n$ that we know the solution from merely knowing the factor.

19. A linear factor of F_u of rank r such as the above will be called a *true* linear factor if $\xi_1, \xi_2, ..., \xi_r$ are independent of $u_1, u_2, ..., u_n$, that is, if it is linear in $x, u_1, u_2, ..., u_n$.

If a linear factor of F_u is not a true linear factor the solution supplied by it is an imbedded one.

Let $x - \xi$ or $x - u_1\xi_1 - ... - u_s\xi_s - u_{s+1}x_{s+1} - ... - u_n x_n$ be a non-true linear factor of F_u, so that $\xi_1, \xi_2, ..., \xi_s$ depend on $u_1, u_2, ..., u_n$. Then $\xi_1, \xi_2, ..., \xi_s, x_{s+1}, ..., x_n$ is a solution of $F_1 = F_2 = ... = F_k = 0$, and so also is $\eta_1, \eta_2, ..., \eta_s, x_{s+1}, ..., x_n$ where $\eta_1, \eta_2, ..., \eta_s$ are obtained from $\xi_1, \xi_2, ..., \xi_s$ by changing $u_1, u_2, ..., u_n$ to $v_1, v_2, ..., v_n$. Hence $\eta, \eta_2,$ ..., $\eta_s, x_{s+1}, ..., x_n$ (where $\eta = u_1\eta_1 + ... + u_s\eta_s + u_{s+1}x_{s+1} + ... + u_n x_n$) is a solution of $f_1 = f_2 = ... = f_k = 0$, and therefore makes F_u vanish. But it does not make $D_u^{(s-1)}...D_u^{(n-1)}$ vanish since this does not involve $x_2, ..., x_s$, and cannot have a factor $x - \eta$, where η involves $v_1, v_2, ..., v_n$. Hence it makes some factor $D_u^{(r-1)}$ of F_u of rank $r < s$ vanish. Then $D_u^{(r-1)}$ vanishes when $x, x_{r+1}, ..., x_s$ are put equal to $\eta, \eta_{r+1}, ..., \eta_s$; and by putting $v_1, v_2, ..., v_n$ (of which $D_u^{(r-1)}$ is independent) equal to $u_1, u_2, ..., u_n$ it follows that $D_u^{(r-1)}$ vanishes when $x, x_{r+1}, ..., x_s$ are put equal to $\xi, \xi_{r+1}, ..., \xi_s$. Hence the solution $\xi, \xi_2, ..., \xi_s, x_{s+1}, ..., x_n$ is an imbedded one (§ 16).

It follows that all the solutions of $F_1 = F_2 = ... = F_k = 0$ are obtainable from true linear factors of F_u; and that all the linear factors of the first complete partial u-resolvent (different from 1) are true linear factors.

It also follows that if there is a spread of rank s which is not imbedded there must be true linear factors of F_u of rank s corresponding to the spread.

We have not proved that all linear factors of F_u are true linear factors[*], and whether this is so or not must be considered doubtful.

20. *If an irreducible factor R_u of F_u considered as a whole function of all the quantities x, x_2, ..., x_n, u_1, u_2, ..., u_n has a true linear factor all its linear factors are true linear factors.*

Let R_u be of rank r. Then R_u is independent of x_1, x_2, ..., x_r and there is a one-one correspondence between its true linear factors and the sets of values ξ_1, ξ_2, ..., ξ_r of $x_1, x_2, ..., x_r$ (not involving $u_1, u_2, ..., u_n$) for which $(R_u)_{x = u_1 x_1 + ... + u_n x_n}$ vanishes. Let

$$(R_u)_{x = u_1 x_1 + ... + u_n x_n} = \rho_1 R_1 + \rho_2 R_2 + ... + \rho_\mu R_\mu,$$

where ρ_1, ρ_2, ..., ρ_μ are different power products of $u_1, u_2, ..., u_n$ and R_1, R_2, ..., R_μ are whole functions of x_1, x_2, ..., x_n independent of $u_1, u_2, ..., u_n$. Then the sets of values ξ_1, ξ_2, ..., ξ_r required are the solutions of $R_1 = R_2 = ... = R_\mu = 0$ regarded as equations for $x_1, x_2, ..., x_r$. These come from the solutions ξ_1, ξ_2, ..., ξ_r, x_{r+1}, ..., x_n of rank r of the same equations in x_1, x_2, ..., x_n.   Now there is at least one solution of rank r, since R_u has a true linear factor ; and only a finite number of such solutions altogether, since R_u has only a finite number of such factors. Hence the first complete partial u-resolvent (different from 1) of the equations $R_1 = R_2 = ... = R_\mu = 0$ is of rank r, and resolves completely into true linear factors (§ 19)

$$x - u_1 \xi_1 - ... - u_r \xi_r - u_{r+1} x_{r+1} - ... - u_n x_n.$$

This complete partial u-resolvent of rank r is therefore R_u itself (or else a power of R_u), which proves the theorem.

If F_u is resolved into factors of the R_u type (irreducible with respect to $x, x_2, ..., x_n, u_1, u_2, ..., u_n$), and these into irreducible factors as regards x, x_2, ..., x_n only, F_u will be resolved into all its irreducible factors. Hence *every irreducible factor of F_u is a factor of a factor of the R_u type, and has all or none of its linear factors true linear factors.*

It follows that any factor of F_u irreducible with respect to x, x_2, ..., x_n, and having a true linear factor, has all its linear factors true linear factors, and is a whole function of $u_1, u_2, ..., u_n$.

---

[*] Kronecker states this as a fact without proving it. König's proof contains an error (K, p. 210). It is not correct to say as he does that $\overline{E}_t^{(\lambda)} \overline{X}_t^{(\lambda)}$ vanishes when $x = \xi_t$, but only when $x, \xi_1, \xi_2, ..., \xi_\lambda$ are put equal to $\xi_t, \xi_1', \xi_2', ..., \xi_\lambda'$.

21. The irreducible spreads of a module. Let R_u be any irreducible factor of F_u of rank r having a true linear factor. We know that

$$R_u = A \prod_{i=1}^{i=d} (x - u_1 x_{1i} - \ldots - u_r x_{ri} - u_{r+1} x_{r+1} - \ldots - u_n x_n).$$

Hence $\quad (R_u)_{x=u_1 x_1 + \ldots + u_n x_n} = A \prod_{i=1}^{i=d} \{u_1 (x_1 - x_{1i}) + \ldots + u_r (x_r - x_{ri})\}.$

To R_u corresponds what is called an *irreducible spread*, viz. the spread of all points $x_{1i}, \ldots, x_{ri}, x_{r+1}, \ldots, x_n$ in which $x_{r+1}, \ldots, x_n$ take all finite values, and $x_{1i}, \ldots, x_{ri}$ the d sets of values supplied by the linear factors of R_u, which vary as $x_{r+1}, \ldots, x_n$ vary.

The degree d of R_u is called the *order* of the irreducible spread.

From the two identities above several useful results can be deduced. It must be remembered that R_u is a known polynomial in x, $x_{r+1}, \ldots,$ x_n, $u_1, u_2, \ldots, u_n$. No linear factor of R_u can be repeated, unless $x_{r+1}, \ldots, x_n$ are given special values; for otherwise R_u and $\frac{\partial R_u}{\partial x}$ would have an H.C.F. involving x, and R_u would be the product of two factors. Whatever set of values $x_{r+1}, \ldots, x_n$ have, whether general or special, the d sets of corresponding values of $x_1, x_2, \ldots, x_r$, viz. $x_{1i}, x_{2i}, \ldots, x_{ri}$ are definite and finite, because R_u is regular in x.

From the second identity it is seen that $(R_u)_{x=u_1 x_1 + \ldots + u_n x_n}$ is independent of $u_{r+1}, \ldots, u_n$, and vanishes identically (i.e. irrespective of $u_1, u_2, \ldots, u_n$) at every point of the spread and no other point. *Hence the whole coefficients* of the power products of $u_1, u_2, \ldots, u_r$ in $(R_u)_{x=u_1 x_1 + \ldots + u_n x_n}$ all vanish at every point of the spread and do not all vanish at any other point.* These coefficients equated to zero give a system of equations for the spread; but it is not necessary to take them all, and some are simpler than others. The coefficient of u_r^d gives an equation $\phi(x_r, x_{r+1}, \ldots, x_n) = A \prod (x_r - x_{ri}) = 0$ for x_r, whose roots are the d values of x_r corresponding to given arbitrary values of $x_{r+1}, \ldots, x_n$. The coefficient of $u_1 u_r^{d-1}$ gives an equation

$$x_1 \phi' - \phi_1 = \phi \Sigma \frac{x_1 - x_{1i}}{x_r - x_{ri}} = 0,$$

where ϕ' is $\frac{\partial \phi}{\partial x_r}$ and ϕ_1, or $\phi \Sigma \frac{x_{1i}}{x_r - x_{ri}}$, is a polynomial in $x_r, x_{r+1}, \ldots, x_n$.

---

* Also these coefficients are members of $(F_1, F_2, \ldots, F_k)$ if $(R_u)_{x=u_1 x_1 + \ldots + u_n x_n}$ is a member of $(F_1, F_2, \ldots, F_k)$, as it will be proved to be when $(F_1, F_2, \ldots, F_k)$ is a prime module (§ 31).

Similarly we have $x_2\phi' - \phi_2 = 0, ..., x_{r-1}\phi' - \phi_{r-1} = 0$.   The equations

$$\phi = 0, \quad x_1 = \frac{\phi_1}{\phi'}, \quad x_2 = \frac{\phi_2}{\phi'}, \quad ..., \quad x_{r-1} = \frac{\phi_{r-1}}{\phi'}$$

are called more particularly the equations of the spread, the first giving the different values of x_r as functions of $x_{r+1}, ..., x_n$, and the others giving $x_1, x_2, ..., x_{r-1}$ as rational functions of $x_r, x_{r+1}, ..., x_n$.   If $x_r, x_{r+1}, ..., x_n$ have such values that $\phi = \phi' = 0$ then $\phi_1, \phi_2, ..., \phi_{r-1}$ all vanish and the expressions above for $x_1, x_2, ..., x_{r-1}$ become indeterminate.   In such a case the values of $x_1, x_2, ..., x_{r-1}$ may be found by taking other equations from $(R_u)_{x = u_1 x_1 + ... + u_n x_n}$ for them.

22. Geometrical property of an irreducible spread.
An algebraic spread in general is one which is determined by any finite system of algebraic equations, and consists of all points whose coordinates satisfy the equations and no other points.   Such a spread has already been shown to consist of a finite number of irreducible spreads each of which is determined by a finite system of equations. The characteristic property of an irreducible spread is that any algebraic spread which contains a part of it, of the same dimensions as the irreducible spread, contains the whole of it.

Let $F_1 = F_2 = ... = F_k = 0$ be the equations determining any algebraic spread, and $F_1' = F_2' = ... = F'_{k'} = 0$ the equations determining an irreducible spread.   The spread they have in common is determined by the combined system of equations $F_1 = F_2 = ... = F_k = F_1' = ... = F''_{k'} = 0$, and is contained in the irreducible spread and has the same or less dimensions.   If it is of the same dimensions as the irreducible spread the complete u-resolvent of $F_1 = ... = F_k = F_1' = ... = F'_{k'} = 0$ will have an irreducible factor R_u'' of the same rank as the irreducible factor R_u' of the complete u-resolvent of $F_1' = F_2' = ... = F'_{k'} = 0$ corresponding to the spread of the same.   Also all the roots of $R_u'' = 0$ regarded as an equation for x are roots of $R_u' = 0$.   Hence R_u' is divisible by R_u'', and since they are both irreducible they must be identical.   Hence the spread of $F_1 = ... = F_k = F_1' = ... = F''_{k'} = 0$ contains the whole of the spread of $F_1' = F_2' = ... = F''_{k'} = 0$, and the spread of $F_1 = F_2 = ... = F_k = 0$ contains the same.   This proves the property stated above.

III. GENERAL PROPERTIES OF MODULES

23. Several arithmetical terms are used in connection with modules suggesting an analogy between the properties of polynomials and the properties of natural numbers. Two modules have a G.C.M., an L.C.M., a product, and a residual (integral quotient); but no sum or difference. Also a prime module answers to a prime number and a primary module to a power of a prime number. Such terms must not be used for making deductions by analogy.

Definitions. Any member F of a module M is said to *contain M*. Also the module (F) contains M. It is immaterial in this statement as in many others whether we regard F as a polynomial or a module. The term *contains* is used as an extension and generalisation of the phrase *is divisible by.*

More generally a module M is said to *contain* another M' if every member of M contains M'; and this will be the case if every member of the basis of M contains M'. Thus $(F_1, F_2, ..., F_k)$ contains $(F_1, F_2, ..., F_{k+1})$, and a module becomes less by adding new members to it.

If M contains M' and M' contains M we say that M, M' are the same module, or $M = M'$.

If M contains M' the spread of M contains the spread of M', but the converse is not true in general.

If in a given finite or infinite set of modules there is one which is contained in every other one, that one is called the *least* module of the set; or if there is one which contains every other one, that one is called the *greatest* module of the set. Two modules cannot be compared as to greater or less unless one contains the other.

There is a module which is contained in all modules, the *unit module* (1). Also (0) may be conceived of as a module which contains all modules; but it seldom comes into consideration and will not be mentioned again. These two modules are called non-proper modules, and all others are *proper* modules. In general by a module a proper module is to be understood.

The G.C.M. of k given modules $M_1, M_2, ..., M_k$ is the greatest of all modules M contained in M_1 and M_2... and M_k, and is denoted by $(M_1, M_2, ..., M_k)$. In order that M may be contained in each of $M_1, M_2, ..., M_k$, or that each of $M_1, M_2, ..., M_k$ may contain M, it is necessary and sufficient that all the members of the bases of M_1, M_2, ..., M_k should contain M; hence the module whose basis consists of all these members contains all the modules M, and is at the same time one of the modules M. It is therefore the greatest of all the modules M and the G.C.M. of M_1, M_2, ..., M_k. The notation $(M_1, M_2, ..., M_k)$ agrees with the notation $(F_1, F_2, ..., F_k)$, since the latter is the G.C.M. of $F_1, F_2, ..., F_k$ regarded as modules.

The L.C.M. of M_1, M_2, ..., M_k is the least of all modules M containing M_1 and M_2 ... and M_k, and is denoted by $[M_1, M_2, ..., M_k]$. Its members consist of all polynomials which contain M_1 and M_2 ... and M_k; for the basis of any module M containing M_1 and M_2 ... and M_k must consist of a certain number of such polynomials, and the whole aggregate of such polynomials constitutes a module M which is the least of all the modules M.

The *product* of M_1, M_2, ..., M_k is the module whose basis consists of all products $F_1 F_2 ... F_k$, where F_i is any member of the basis of M_i ($i = 1, 2, ..., k$). The product is denoted by $M_1 M_2 ... M_k$, and is evidently a definite module independent of what bases may be chosen for M_1, M_2, ..., M_k. The product $M_1 M_2 ... M_k$ contains the L.C.M. $[M_1, M_2, ..., M_k]$.

The product of γ modules each of which is the same module M is denoted by M^γ and is called a power of M. If P is the point $(a_1, a_2, ..., a_n)$ the module $(x_1 - a_1, x_2 - a_2, ..., x_n - a_n)$ is denoted by P. If O is the origin the module O is $(x_1, x_2, ..., x_n)$, and O^γ is a module having for basis all power products of $x_1, x_2, ..., x_n$ of degree γ. A polynomial F, or module M, which contains P^γ is said to have a γ-point at P.

The *residual* (L, p. 49) of a given module M' with respect to another M is the least module whose product with M' contains M and is denoted by M/M'. Its members consist of every polynomial whose product with each member separately of the basis of M' is a member of M; for the basis of any module whose product with M' contains M must consist of a certain number of such polynomials, and the whole aggregate of such polynomials constitutes the least such module.

In the case of the natural numbers the residual of m' with respect to m is the least number whose product with m' contains m, and is the quotient of m by the G.C.M. of m and m'. It is the same to some extent with modules, viz. $M/M' = M/(M, M')$; for if $M/M' = M''$ then M'' is the least module such that $M'M''$ contains M, and is therefore the least module such that $(M, M')M''$ contains M, i.e. $M'' = M/(M, M')$.

Nevertheless $M/(M, M')$ is not called the quotient of M by (M, M') because it is not true in general that the product of (M, M') and $M/(M, M')$ is M.

If M, M', M'' are three modules such that $M'M''$ contains M it is clear that M' contains M/M'' and M'' contains M/M'. Since MM' contains M, M contains M/M'. The module M/M' is a module contained in M having a special relation to M independently of what M' may be (§ 26 (i)).

There is a least module which can be substituted for M' without changing M/M', viz. $M/(M/M')$, § 26 (ii). This module is contained in (M, M'), for (M, M') can be substituted for M' without changing M/M', but is in general different from (M, M').

24. Comment on the definitions. The non-proper unit module (1) has no spread. Conversely a module which has no spread is the module (1), since the complete resolvent is 1 and is a member of the module. The unit module is of importance from the fact that it often comes at the end of a series of modules derived by some process from a given module.

$(M_1, M_2, ..., M_k)$ and $[M_1, M_2, ..., M_k]$ obey the associative law $[M_1, M_2, M_3] = [[M_1, M_2], M_3] = [M_1, [M_2, M_3]]$, and the commutative law $(M_1, M_2) = (M_2, M_1)$. Also $(M_1, M_2, ..., M_k)$ obeys the distributive law $M(M_1, M_2) = (MM_1, MM_2)$; but $[M_1, M_2, ..., M_k]$ does not.

Example. As an example of the last statement we have

$$(x_1, x_2)[(x_1^2, x_2^2), (x_1 x_2)] = (x_1, x_2)(x_1^2 x_2, x_1 x_2^2) = (x_1 x_2)(x_1, x_2)^2,$$

while $\qquad [(x_1, x_2)(x_1^2, x_2^2), (x_1, x_2)(x_1 x_2)] = (x_1 x_2)(x_1, x_2).$

Given the bases of $M_1, M_2, ..., M_k, M, M'$ we know at once a basis for $(M_1, M_2, ..., M_k)$ and for $M_1 M_2 ... M_k$; but it may be extremely difficult to find a basis for $[M_1, M_2, ..., M_k]$ or for M/M'. Hilbert (H, pp. 492-4, 517) has given a process for finding a basis of $[M_1, M_2, ..., M_k]$; and the same process can be applied for finding a basis for M/M'. This process is chiefly of theoretical value in so far as it has any value.

We can have (i) $MM' = MM''$, or $M/M' = M/M''$, without $M' = M''$; (ii) $M/M' = M''$ without $M/M'' = M'$; (iii) $M/M' = M''$ and $M/M'' = M'$ without $M = M'M''$; and (iv) $M = M'M''$ without $M/M' = M''$ or $M/M'' = M'$.

Examples. (i) $(x_1, x_2)(x_1, x_2)^2 = (x_1, x_2)(x_1^2, x_2^2),$
$(x_1, x_2)^3/(x_1, x_2)^2 = (x_1, x_2)^3/(x_1^2, x_2^2);$

(ii) $(x_1, x_2)^3/(x_1^2, x_2^2) = (x_1, x_2)$, while $(x_1, x_2)^3/(x_1, x_2) = (x_1, x_2)^2$;

(iii) $(x_1^2, x_2^2)/(x_1, x_2) = (x_1, x_2)^2$ and $(x_1^2, x_2^2)/(x_1, x_2)^2 = (x_1, x_2)$,
while $(x_1^2, x_2^2) \neq (x_1, x_2)(x_1, x_2)^2$;

(iv) $(x_1, x_2)^6 = (x_1^3, x_1^2 x_2, x_2^3)(x_1^3, x_1 x_2^2, x_2^3)$,
while $(x_1, x_2)^6/(x_1^3, x_1^2 x_2, x_2^3)$ and $(x_1, x_2)^6/(x_1^3, x_1 x_2^2, x_2^3)$ are both
equal to $(x_1, x_2)^3$.

25. *The product of the* G.C.M. *and* L.C.M. *of two modules contains
the product of the modules.*

Let $M = (F_1, F_2, ..., F_k)$ and $M' = (F_1', F_2', ..., F'_\kappa)$ be the two
modules and let F_L be any member of the basis of their L.C.M. Then,
since $F_L = 0 \bmod M$, $F_i' F_L = 0 \bmod MM'$; and since $F_L = 0 \bmod M'$,
$F_i F_L = 0 \bmod MM'$; i.e. the product of any member of the basis of
(M, M') with any member of the basis of $[M, M']$ contains MM', or
$(M, M')[M, M']$ contains MM'.

When M, M' have no point in common $(M, M') = (1)$ and consequently $[M, M']$ contains MM', i.e. $[M, M'] = MM'$. This case is
proved by König (K, p. 356); although it is to be noticed that
(M, M') cannot be (1) in the case of modules of homogeneous polynomials. Thus the L.C.M. of any finite number of simple modules
(§ 33) is the same as their product (Mo).

26. *The modules M/M' and $M/(M/M')$ are mutually residual
with respect to M, i.e. each is the residual of the other with respect to M.*

Let $M/M' = M''$ and $M/(M/M') = M'''$; then we have $M''' = M/M''$,
and we have to prove that $M'' = M/M'''$. Let $M/M''' = M^{iv}$. Now
$M'M''$ contains M; therefore M' contains M/M'' or M'''. Also
$M''M'''$ contains M; therefore M'' contains M/M''' or M^{iv}. Again,
since M' contains M''' (proved) and $M'''M^{iv}$ contains M, $M'M^{iv}$
contains M, i.e. M^{iv} contains M/M' or M''. But M'' contains M^{iv}
(proved). Hence $M'' = M^{iv} = M/M'''$.

Two results follow from this :

(i) *M/M' is a module contained in M of a particular type ;* for
M/M' and its residual with respect to M are mutually residual with
respect to M, and this is not true in general of any module contained
in M and the residual module (Ex. ii, § 24).

(ii) *The least module which can be substituted for M' without
changing M/M' is $M/(M/M')$.* Let M^{iv} be any module such that
$M/M^{iv} = M/M'$; then the product of M^{iv} and M/M' contains M, and
M^{iv} contains $M/(M/M')$. Also $M/(M/M')$ is one of the modules M^{iv};

for if $M/(M/M') = M'''$ then $M/M''' = M/M'$, by the theorem. Hence $M/(M/M')$ is the least of the modules M^{iv} which can be substituted for M' without changing M/M'.

27. *If M', M'' are mutually residual with respect to any module they are mutually residual with respect to $M'M''$.*

Suppose M', M'' are mutually residual with respect to M. Then $M'M''$ contains M; and if $M'M''/M' = M'''$, $M'M''$ contains $M'M''$ which contains M; hence M''' contains M/M' or M''. Also M'' contains $M'M''/M'$ or M'''. Hence $M'' = M''' = M'M''/M'$. Similarly $M' = M'M''/M''$ (cf. statement iv, § 24).

Any module M with respect to which M', M'' are mutually residual contains $[M', M'']$ and is contained in $M'M''$.

28. *If $M, M_1, M_2, ..., M_k$ are any modules, then*
$$M/(M_1, M_2, ..., M_k) = [M/M_1, M/M_2, ..., M/M_k],$$
and $\quad [M_1, M_2, ..., M_k]/M = [M_1/M, M_2/M, ..., M_k/M].$

For $M/(M_1, M_2, ..., M_k)$ contains M/M_i and therefore contains $[M/M_1, M/M_2, ..., M/M_k]$. Also $M_i [M/M_1, ..., M/M_k]$ contains $M_i \times M/M_i$ which contains M; hence $(M_1, ..., M_k) [M/M_1, ..., M/M_k]$ contains M, and $[M/M_1, ..., M/M_k]$ contains $M/(M_1, ..., M_k)$. This proves the first part.

Again $[M_1, ..., M_k]/M$ contains M_i/M and therefore contains $[M_1/M, ..., M_k/M]$. Also $M[M_1/M, ..., M_k/M]$ contains M_i and therefore contains $[M_1, ..., M_k]$; hence $[M_1/M, ..., M_k/M]$ contains $[M_1, ..., M_k]/M$. This proves the second part.

29. Prime and Primary Modules. *Definitions.* A *prime module* is defined by the property that no product of two modules contains it without one of them containing it.

A *primary module* is defined by the property that no product of two modules contains it without one of them containing it or both containing its spread. Hence if one does not contain the spread the other contains the module.

Primary modules will be understood to include prime modules.

Lasker introduced and defined the term primary (L, p. 51), though not in the same words as given here. The conception of a primary module is a fundamental one in the theory of modular systems.

Any irreducible spread determines a prime module, viz. the module whose members consist of all polynomials containing the spread. That this module is prime follows from the fact that no product of two polynomials can contain the spread without one of them containing it (§ 22) and the module; and the same is true if for polynomials we write modules.

If $M = (F_1, F_2, ..., F_k)$ is the prime module of rank r determined by an irreducible spread of dimensions $n - r$, and if the origin be moved to a general point of the spread, the constant terms of $F_1, F_2, ..., F_k$ will vanish, and the linear terms will be equivalent to r independent linear polynomials, i.e. the sub-determinants of order $> r$ of the matrix

$$\begin{vmatrix} \dfrac{\partial F_1}{\partial x_1}, & \dfrac{\partial F_1}{\partial x_2}, & \cdots & \dfrac{\partial F_1}{\partial x_n} \\ \cdots\cdots\cdots\cdots\cdots\cdots \\ \dfrac{\partial F_k}{\partial x_1}, & \dfrac{\partial F_k}{\partial x_2}, & \cdots & \dfrac{\partial F_k}{\partial x_n} \end{vmatrix}$$

will vanish, while those of order r will not vanish, at the origin. This will be equally true for any general point of the spread without moving the origin to it. Any point of the spread for which the sub-determinants of order r of this matrix vanish is called a *singular point* of the spread, and the aggregate of such points the *singular spread* contained in the given spread. The singular spread (if any exists) is therefore the spread determined by $F_1, F_2, ..., F_k$ and the sub-determinants of order r of the above matrix.

If $M = (F_1, F_2, ..., F_k)$ is the L.C.M. of the prime modules determined by any finite number of irreducible spreads of the same dimensions $n - r$, the same definition holds concerning singularities of the whole spread. In this case the singular spread consists of the intersections of all pairs of the irreducible spreads, together with all the singular spreads contained in the irreducible spreads considered individually.

30. *The spread of any prime or primary module is irreducible.* For if not the complete u-resolvent has at least two factors corresponding to two different irreducible spreads of the module neither of which contains the other, and is the product of two polynomials neither of which contains the whole spread of the module, i.e. the module is neither prime nor primary.

31. *There is only one prime module with a given (irreducible) spread, viz. the module whose members consist of all polynomials containing the spread.*

Let $M = (F_1, F_2, ..., F_k)$ be any prime module of rank r. It will be sufficient to prove that every polynomial which contains the spread of M contains the module M. The first complete partial u-resolvent of M other than 1 will be a power $R_u{}^m$ of an irreducible polynomial R_u in $x, x_{r+1}, ..., x_n$. Also the complete u-resolvent is a member of $(f_1, f_2, ..., f_k)$, § 18, which is prime; and every factor except $R_u{}^m$ is of too high rank to contain the spread of $(f_1, f_2, ..., f_k)$. Hence $R_u{}^m$, and therefore R_u itself, is a member of $(f_1, f_2, ..., f_k)$. Hence $(R_u)_{x=u_1x_1+...+u_nx_n}$ is a member of M, and also the whole coefficient of any power product of $u_1, u_2, ..., u_n$ in $(R_u)_{x=u_1x_1+...+u_nx_n}$. We have proved (§ 21) that

$$(R_u)_{x=u_1x_1+...+u_nx_n} = ... + u_r{}^{d-1}(u_1\psi_1 + ... + u_{r-1}\psi_{r-1}) + u_r{}^d \phi,$$

where $\psi_1 = x_1\phi' - \phi_1, ..., \psi_{r-1} = x_{r-1}\phi' - \phi_{r-1}$. Hence $\psi_1, ..., \psi_{r-1}, \phi$ are all members of M.

Let F be any polynomial which contains the spread of M. In F put $x_1 = \phi_1/\phi', x_2 = \phi_2/\phi', ..., x_{r-1} = \phi_{r-1}/\phi'$; then F becomes a rational function of $x_r, x_{r+1}, ..., x_n$ of which the denominator is ϕ'^l, where l is the degree of F. This rational function vanishes for all points of the spread at which ϕ' does not vanish, and its numerator is therefore divisible by ϕ. We have then

$$F\left(\frac{\phi_1}{\phi'}, \frac{\phi_2}{\phi'}, ..., \frac{\phi_{r-1}}{\phi'}, x_{r+1}, ..., x_n\right) = \frac{X\phi}{\phi'^l},$$

where X is a whole function of $x_r, x_{r+1}, ..., x_n$; i.e.

$$F\left(x_1 - \frac{\psi_1}{\phi'}, x_2 - \frac{\psi_2}{\phi'}, ..., x_{r-1} - \frac{\psi_{r-1}}{\phi'}, x_{r+1}, ..., x_n\right) = \frac{X\phi}{\phi'^l},$$

or $\phi'^l F(x_1, x_2, ..., x_n) = 0 \mod (\psi_1, ..., \psi_{r-1}, \phi) = 0 \mod M$.
Hence $F = 0 \mod M$, which proves the theorem.

It follows that a module which is the L.C.M. of a finite number of prime modules, whether of the same rank or not, is uniquely determined by its spread, and any polynomial containing the spread contains the module.

32. *If M is a primary module and M_1 the prime module determined by its spread some finite power of M_1 contains M.*

This theorem, in conjunction with Lasker's theorem (§ 39), is equivalent to the Hilbert-Netto theorem (§ 46). The proofs of the theorem by Lasker and König are both wrong. Lasker first assumes the theorem (L, p. 51) and then proves it (L, p. 56); and König makes an absurdly false assumption concerning divisibility (K, p. 399).

By the same reasoning as in the last theorem it follows that $R_u{}^m$ (but not R_u) is a member of $(f_1, f_2, \ldots, f_k)$, and

$$(R_u{}^m)_{x=u_1 x_1 + \ldots + u_n x_n} = \{\ldots + u_r{}^{d-1}(u_1 \psi_1 + \ldots + u_{r-1} \psi_{r-1}) + u_r{}^d \phi\}^m = 0 \bmod M.$$

Picking out the coefficients of $u_r{}^{dm}$ and $u_r{}^{dm-m} u_1{}^m$, we have

$$\phi^m = 0 \bmod M, \quad \text{and} \quad \psi_1{}^m = X\phi \bmod M; \quad \therefore \quad \psi_1{}^{m^2} = 0 \bmod M;$$

and similarly $\psi_2{}^{m^2} = \ldots = \psi_{r-1}{}^{m^2} = 0 \bmod M$. Also if F is any member of M_1, then, by the last theorem,

$$\phi^{'l} F = 0 \bmod (\psi_1, \ldots, \psi_{r-1}, \phi).$$

Hence the product of any rm^2 polynomials F and $\phi^{'lrm^2}$ is a member of $(\psi_1{}^{m^2}, \psi_2, \ldots \psi_{r-1}{}^{m^2}, \phi^{m^2})$ and of M, i.e. $M_1{}^{rm^2}$ contains M.

33. *Definitions.* If M is a primary module and M_1 the corresponding prime module the least number γ such that $M_1{}^\gamma$ contains M is called the *characteristic number* of M.

A *simple* module is a module containing one point only (Mo). For example, $O^\gamma = (x_1, x_2, \ldots, x_n)^\gamma$ is a simple module with characteristic number γ.

A module of homogeneous polynomials will be called an *H-module*. A simple *H*-module has the origin for its spread; but a simple module having the origin for spread is not in general an *H*-module.

A simple module is primary. For if M is a simple module, and M', M'' any two modules whose product contains M, of which M' does not contain the spread of M, then (M, M') contains no point and is the module (1); but $(M, M') M''$ contains M, i.e. M'' contains M; hence M is primary.

34. *There is no higher limit to the number of members that may be required to constitute a basis of a prime module.* This is not in conflict with Kronecker's statement, proved by König (K, p. 234), that there always exist $n + 1$ polynomials containing a given algebraic spread which have no point in common outside the spread.

Example. Consider $\frac{1}{2} l(l-1)$ straight lines through the origin O in 3-dimensional space, not lying on any cone of order $l-2$. Draw a cone of order l and a surface (not a cone) of order l through the $\frac{1}{2} l(l-1)$ lines so as to intersect again in an irreducible curve of order $\frac{1}{2} l(l+1)$ with $\frac{1}{2} l(l-1)$ tangents at O. Then no basis of the prime module determined by this curve can have less than l members, where l is a number which can be chosen as high as we please.

This can be proved by considering residuation on the cone. The original $\frac{1}{2} l(l-1)$ generators have a residual on the cone of $\frac{1}{2} l(l-1)$ generators, which again have a residual of $\frac{1}{2} l(l-1)$ generators, of which $l-1$ can be chosen at will. This last set of generators is residual to the irreducible curve and together they make the whole intersection of the cone with a surface of order l having an $(l-1)$-point at O. Hence there are l surfaces of order l containing the irreducible curve which have an $(l-1)$-point at O and in which the terms of degree $l-1$ are linearly independent, while there is no surface containing the curve with less than an $(l-1)$-point at O. The prime module determined by the curve must therefore have at least l members in its basis. The module has in fact a basis of $l+1$ members, the $l+1$ linearly independent surfaces of order l containing it (including the cone); and these can be reduced to l members.

In the case $n = 2$ the curve is an ordinary space cubic determining a prime module

$$(f_1, f_2, f_3) = (vw' - v'w, \; wu' - w'u, \; uv' - u'v),$$

where u, v, w, u', v', w' are linear. The basis of three members can be reduced to two $f_1 - af_2, f_1 - bf_3$ provided constants a, b, λ, λ' and linear functions α, β can be chosen so that

$$f_1 = \alpha \, (f_1 - af_2) + \beta \, (f_1 - bf_3),$$

or $\qquad\qquad (1 - \alpha - \beta) f_1 + a\alpha f_2 + b\beta f_3 = 0,$

or $\qquad 1 - \alpha - \beta = \lambda u + \lambda' u', \quad a\alpha = \lambda v + \lambda' v', \quad b\beta = \lambda w + \lambda' w' \, ;$

and this can be done.

35. *The* L.C.M. *of any number of primary modules with the same spread is a primary module with the same spread.*

Let $M_1, M_2, \ldots, M_k$ be primary modules with the same spread, and let M be their L.C.M. Then M has the same irreducible spread, since the product, which contains the L.C.M., has the same spread. Also if the product $M'M''$ contains M, and M' does not contain the spread, then M'' contains M_1 and $M_2 \ldots$ and M_k, i.e. M'' contains M. Hence M is primary. The G.C.M. is not primary in general.

36. *If M is primary and M' is any module not containing M then M/M' is primary and has the same spread as M.*

Let $M/M' = M''$. Then since $M'M''$ contains M, and M' does not contain M, M'' contains the spread of M. Also M contains M''; hence M'' has the same spread as M. Also if $M_1 M_2$ contains M'' then

$M'M_1M_2$ contains $M'M''$ which contains M; and if M_1 does not contain the spread of M (that is of M''') $M'M_2$ contains M, and M_2 contains M/M' or M''; i.e. M'' is primary.

37. Hilbert's Theorem (H, p. 474). *If F_1, F_2, F_3, ... is an infinite series of homogeneous polynomials there exists a finite number k such that $F_h = 0$ mod $(F_1, F_2, ..., F_k)$ when $h > k$.*

The following proof is substantially König's (K, p. 362). It must be clearly understood that $F_1, F_2, F_3, ...$ *are given in a definite order.* In the case of a single variable the series F_1, F_2, F_3, ... consists of powers of the variable, and if F_k is the least power then $F_h = 0$ mod F_k when $h > k$. Hence the theorem is true in this case. We shall assume it for $n - 1$ variables and prove it for n variables.

The series F_1', F_2', F_3', ... is called a modified form of the series $F_1, F_2, F_3, ...$ if $F_1' = F_1$ and $F_i' = F_i$ mod $(F_1, F_2, ..., F_{i-1})$ for $i > 1$. Thus the modules $(F_1, F_2, ..., F_i)$ and $(F_1', F_2', ..., F_i')$ are the same. The theorem will be proved if we show that the series F_1', F_2', ... can be so chosen that all its terms after a certain finite number become zero. We assume that F_1 is regular in x_n, and we choose the modified series so that each of its terms F_i' after the first is of as low degree as possible in x_n, and therefore of lower degree in x_n than F_1'. The terms of the series F_1', F_2', ... of degree zero in x_n will be polynomials in $x_1, x_2, .., x_{n-1}$ and these can be modified so that all after a certain finite number become zero, since the theorem is assumed true for $n - 1$ variables. Let $F'_{l_1}, F'_{l_2}, F'_{l_3}, ...$ be all the terms of $F_1', F_2', F_3', ...,$ taken in order, which are of one and the same degree $l > 0$ in x_n; and let $f'_{l_1}, f'_{l_2}, ...$ be the whole coefficients of x_n^l in them. Then $f'_{l_1}, f'_{l_2}, f'_{l_3}, ...$ are polynomials in $n - 1$ variables; and we cannot have $f'_{l_i} = 0$ mod $(f'_{l_1}, f'_{l_2}, .., f'_{l_{i-1}})$ for any value of i; for if $f'_{l_i} = A_1 f'_{l_1} + A_2 f'_{l_2} + \cdots + A_{i-1} f'_{l_{i-1}}$, then $F'_{l_i} - A_1 F'_{l_1} - \cdots - A_{i-1} F'_{l_{i-1}}$ is of less degree than l in x_n, which cannot be. Hence the number of the polynomials $f'_{l_1}, f'_{l_2}, ...,$ or the number of terms $F'_{l_1}, F'_{l_2}, ...$ in the series $F_1', F_2', ...,$ is finite. And the number of values of l is also finite, the greatest value of l being the value it has in F_1'. Hence the theorem is proved.

The theorem can be extended at once to an infinite series $F_1, F_2, ...$ of non-homogeneous polynomials since they can all be made homogeneous by introducing a variable x_0 of homogeneity.

The following is an immediate consequence of the theorem :

Any module of polynomials has a basis consisting of a finite number of members.

To prove this it is only necessary to show that a complete linearly independent set of members of any module can be arranged in a definite order in an infinite series. If l is the lowest degree of any member we can first take any complete linearly independent set of members of degree l, then any complete set of members of degree $l + 1$ whose terms of degree $l + 1$ are linearly independent, then a similar set of members of degree $l + 2$, and so on. In this way a complete linearly independent set of members is obtained in a definite order. It does not matter in what order the members of a set are taken, nor is it necessary to know how to find the members of a set. It is sufficient to know that there is a definite finite number of members belonging to each set.

38. The H-module equivalent to a given module. Consider a complete linearly independent set of members of a given module M, not an H-module, arranged in a series in the order described above; and make all the members homogeneous by introducing a new variable x_0. We then have a series of homogeneous polynomials belonging to an H-module M_0, whose basis consists of a finite number of members of the series. The module M_0 is called *the H-module equivalent to M*, and a basis of M obtained from any basis of M_0 by putting $x_0 = 1$ is called an *H-basis* of M. The distinctive property of an H-basis $(F_1, F_2, ..., F_k)$ of M is that any member F of M can be put in the form $A_1F_1 + A_2F_2 + ... + A_kF_k$ where A_iF_i $(i = 1, 2, ..., k)$ *is not of greater degree than F*. *Every module has an H-basis*, which may necessarily consist of more members than would suffice for a basis in general.

The following relations exist between M and its equivalent H-module M_0: (i) to any member F of M corresponds a member F_0 of M_0 of the same degree as F, and an infinity of members $x_0^p F_0$ of higher degree; (ii) to any member F_0 of M_0 corresponds one and only one member of M, viz. $(F_0)_{x_0=1}$; (iii) there is a one-one correspondence between the members of M_0 of degree l and the members of M of degree $\leqslant l$.

If $x_0 F_0 = 0 \bmod M_0$, then $(F_0)_{x_0=1} = 0 \bmod M$, and $F_0 = 0 \bmod M_0$ by (i), i.e. there is no member $x_0 F_0$ of M_0 such that F_0 is not a member of M_0, and $M_0/(x_0) = M_0$. Conversely *an H-module M in n variables $x_1, x_2, ..., x_n$ is equivalent to the module $M_{x_n=1}$ if $M/(x_n) = M$, and not otherwise.*

In any basis $(F_1, F_2, \ldots, F_k)$ of an H-module in which no member is irrelevant, i.e. no $F_i = 0 \bmod (F_1, \ldots, F_{i-1}, F_{i+1}, \ldots, F_k)$, the number of members of each degree is fixed; as can be easily seen by arranging $F_1, F_2, \ldots, F_k$ in order of degree. *Hence in any H-basis of a module in which no member is irrelevant the number of members of each degree is fixed.* On account of this and the other properties of an H-basis mentioned above an H-basis gives a simpler and clearer representation of a module than a basis which is not an H-basis.

Example. Find an H-basis of the module $(x_1^2, x_2 + x_1 x_3)$.

Take the H-module $(x_1^2, x_2 x_0 + x_1 x_3)$ and solve the equation

$$x_0 X_0 = 0 \bmod (x_1^2, x_2 x_0 + x_1 x_3),$$

or $\qquad\qquad x_0 X_0 = x_1^2 X_1 + (x_2 x_0 + x_1 x_3) X_2.$

Putting $x_0 = 0$ we have

$$(x_1^2 X_1 + x_1 x_3 X_2)_{x_0 = 0} = 0,$$

i.e. $\qquad\qquad X_1 = x_3 X, \quad X_2 = - x_1 X, \quad \text{when } x_0 = 0,$

i.e. $\qquad\qquad X_1 = x_3 X + x_0 Y_1, \quad X_2 = - x_1 X + x_0 Y_2.$

Hence

$$x_0 X_0 = x_1^2 (x_3 X + x_0 Y_1) + (x_2 x_0 + x_1 x_3)(- x_1 X + x_0 Y_2)$$
$$= x_0 (x_1^2 Y_1 - x_1 x_2 X + \overline{x_2 x_0 + x_1 x_3}\, Y_2),$$

i.e. $\qquad\qquad X_0 = 0 \bmod (x_1^2, x_1 x_2, x_2 x_0 + x_1 x_3).$

Again, if we solve the equation

$$x_0 Y_0 = 0 \bmod (x_1^2, x_1 x_2, x_2 x_0 + x_1 x_3),$$

we find $\qquad\qquad Y_0 = 0 \bmod (x_1^2, x_1 x_2, x_2^2, x_2 x_0 + x_1 x_3);$

and if we solve

$$x_0 Z_0 = 0 \bmod (x_1^2, x_1 x_2, x_2^2, x_2 x_0 + x_1 x_3),$$

we find $\qquad\qquad Z_0 = 0 \bmod (x_1^2, x_1 x_2, x_2^2, x_2 x_0 + x_1 x_3).$

Hence $(x_1^2, x_1 x_2, x_2^2, x_2 x_0 + x_1 x_3)$ is the H-module equivalent to $(x_1^2, x_2 + x_1 x_3)$, and $(x_1^2, x_1 x_2, x_2^2, x_2 + x_1 x_3)$ is an H-basis of $(x_1^2, x_2 + x_1 x_3)$.

The extra members $x_1 x_2, x_2^2$ might of course have been found more quickly by multiplying $x_2 + x_1 x_3$ first by x_1 and then by x_2. The method given is a general one.

39. Lasker's Theorem (L, p. 51). *Any given module M is the* L.C.M. *of a finite number of primary modules.*

Let M be of rank r. Express its first complete partial u-resolvent $D_u^{(r-1)}$ in irreducible factors, viz.

$$D_u^{(r-1)} = R_1^{m_1} R_2^{m_2} \ldots R_j^{m_j};$$

and let C_1, C_2, ..., C_j denote the irreducible spreads, of dimensions $n-r$, corresponding to R_1, R_2, ..., R_j respectively.

Consider the whole aggregate M_i of polynomials F for each of which there exists a polynomial F', not containing C_i, such that $FF'=0 \bmod M$. We shall prove first that M_i is a primary module whose spread is C_i $(i = 1, 2, ..., j)$.

Let F_1, F_2 be any two members of M_i. Then since $F_1F_1' = 0 \bmod M$, and $F_2F_2' = 0 \bmod M$, where neither F_1' nor F_2' contains C_i, we have $(A_1F_1 + A_2F_2)F_1'F_2' = 0 \bmod M$, where $F_1'F_2'$ does not contain C_i. Hence $A_1F_1 + A_2F_2$ belongs to the aggregate M_i, i.e. M_i is a module.

Again, since $FF' = 0 \bmod M$, F contains C_i, and M_i contains C_i. Now, if F_u is the complete u-resolvent of M,

$$(F_u)_{x=u_1x_1+...+u_nx_n} = 0 \bmod M,$$

while $(R_i^{m_i})_{x=u_1x_1+...+u_nx_n}$ is the only factor of $(F_u)_{x=u_1x_1+...+u_nx_n}$ which contains C_i. Hence $(R_i^{m_i})_{x=u_1x_1+...+u_nx_n} = 0 \bmod M_i$. But the polynomial $(R_i)_{x=u_1x_1+...+u_nx_n}$ does not vanish identically (i.e. irrespective of $u_1, u_2, ..., u_n$) for any point outside C_i (§ 21); hence M_i contains no point outside C_i, i.e. C_i is the spread of M_i.

Lastly M_i is primary; for if $F''F''' = 0 \bmod M_i$, then

$$F'F''F''' = 0 \bmod M,$$

where F' does not contain C_i; hence, if F'' does not contain C_i, $F'F''$ does not, and $F''' = 0 \bmod M_i$. Hence also if $M''M'''$ contains M_i, and M'' does not contain C_i, M''' contains M_i. Thus M_i is a primary module whose spread is C_i. Also M contains M_i, for every member of M is a member of M_i.

The module M/M_i does not contain C_i; for if $M_i = (F_1, F_2, ..., F_k)$ and F_1', F_2', ..., F_k' are polynomials not containing C_i such that

$$F_lF_l' = 0 \bmod M \quad (l = 1, 2, ..., k),$$

then $\qquad F_lF_1'F_2'...F_k' = 0 \bmod M \quad (l = 1, 2, ..., k).$

Hence $F_1'F_2'...F_k'$ is a member of M/M_i not containing C_i; and therefore M/M_i cannot contain C_i.

Since M/M_i does not contain C_i, $(M/M_1, M/M_2, ..., M/M_j)$ does not contain any of the spreads C_1, C_2, ..., C_j. We can now prove that if ϕ is any single member of $(M/M_1, M/M_2, ..., M/M_j)$ which does not contain any of the spreads C_1, C_2, ..., C_j, then

$$M = [M_1, M_2, ..., M_j, (M, \phi)].$$

Since M contains $[M_1, M_2, ..., M_j, (M, \phi)]$ it has only to be proved that the latter contains M, or that

$$F = 0 \bmod [M_1, M_2, ..., M_j, (M, \phi)] \text{ requires } F = 0 \bmod M.$$

We have $F = 0 \bmod (M, \phi) = f\phi \bmod M = f\phi \bmod M_i$;

but $F = 0 \bmod M_i$; therefore $f\phi = 0 \bmod M_i$, and, since ϕ does not contain C_i,

$$f = 0 \bmod M_i = 0 \bmod [M_1, M_2, ..., M_j].$$

Hence $f\phi = 0 \bmod [M_1, M_2, ..., M_j] (M/M_1, ..., M/M_j) = 0 \bmod M$ (§ 28), and $F = f\phi \bmod M = 0 \bmod M$. Hence $M = [M_1, M_2, ..., M_j, (M, \phi)]$.

Now the spread of (M, ϕ) is of dimensions $< n - r$, since ϕ does not contain any spread of M of dimensions $n - r$. Hence the same process can be applied to (M, ϕ) as to M; and we finally arrive at a module $(M, \phi, \phi', ...)$ with no spread, which is the module (1). Hence $M = [Q_1, Q_2, ..., Q_k]$ where $Q_1, Q_2, ..., Q_k$ are all primary modules of ranks $\leqslant r$.

40. Comment on Lasker's Theorem. The above is in all essentials the remarkable proof given by Lasker of this fundamental theorem. He considers H-modules only and makes use of homogeneous coordinates, in consequence of which his enunciation of the theorem is not quite as simple as the one above.

Any module among $Q_1, Q_2, ..., Q_k$ which is contained in the L.C.M. of all the rest is *irrelevant* and may be omitted. It will be understood in writing $M = [Q_1, Q_2, ..., Q_k]$ that all irrelevant modules have been omitted. Those that remain will be called the *relevant primary modules* into which M resolves, and their spreads will be called the *relevant spreads* of M. A relevant spread which is not contained in another of higher dimensions is called an *isolated spread* and the corresponding module an *isolated primary module* of M. The other relevant spreads and modules are called *imbedded* spreads and modules of M. All the relevant spreads of M whether isolated or imbedded are unique. Also the isolated primary modules are unique, but the imbedded primary modules are to some extent indeterminate.

A process by which $Q_1, Q_2, ..., Q_k$ can be theoretically obtained, without bringing in any irrelevant modules, is described in (M). The isolated spreads are found from the irreducible factors of the complete u-resolvent after rejecting all factors which give imbedded spreads. To these correspond unique primary modules of M which can be found. Let $M^{(0)}$ be their L.C.M. The isolated spreads of $M/M^{(0)}$ are the relevant spreads of M imbedded to the first degree. To these correspond indeterminate imbedded primary modules of M which are chosen as simply as possible. Although not uniquely determinate the L.C.M. of each one and $M^{(0)}$ is unique, and the L.C.M. of them all and $M^{(0)}$ is a unique module $M^{(1)}$. The isolated spreads of $M/M^{(1)}$ are the relevant spreads of M imbedded to the second degree; and the L.C.M. of the corresponding (indeterminate) primary modules and $M^{(1)}$ is a unique module $M^{(2)}$. The process is continued until a module $M^{(t)}$ is obtained such that $M/M^{(t)} = (1)$, when there will be no more relevant primary modules to find.

41. An *unmixed module* is usually understood to be one whose *isolated* irreducible spreads are all of the same dimensions; but it is clear from the above that this cannot be regarded as a satisfactory view. It should be defined as follows:

Definition. An *unmixed module* is one whose relevant spreads, both isolated and imbedded, are all of the same dimensions; and a *mixed module* is one having at least two relevant spreads of different dimensions.

An unmixed module cannot have any relevant imbedded spreads.

A primary module is an unmixed module whose spread is irreducible. This cannot be taken as a definition because the meaning of *unmixed* depends on the meaning of *primary*.

Condition that a module may be unmixed. In order that a module M of rank r may be unmixed it is necessary and sufficient that it should have no relevant spread of rank $> r$. This condition may be expressed by saying that $\phi F = 0 \bmod M$ requires $F = 0 \bmod M$ where ϕ is any polynomial involving $x_{r+1}, \ldots, x_n$ only. For if M contains a relevant primary module of rank $> r$ a ϕ can be chosen which contains it, and an F which does not contain it but contains all the other relevant primary modules of M, so that $\phi F = 0 \bmod M$ does not require $F = 0 \bmod M$; while if M contains no relevant primary module of rank $> r$ there is no ϕ containing a relevant spread of M and $\phi F = 0 \bmod M$ requires $F = 0 \bmod M/(\phi) = 0 \bmod M$ (§ 42).

A primary module Q has a certain *multiplicity* (§ 68). To a given primary module $Q^{(\mu)}$ of multiplicity μ corresponds a series of primary modules $Q^{(1)}, Q^{(2)}, \ldots, Q^{(\mu)}$ of multiplicities $1, 2, \ldots, \mu$ all having the same spread as $Q^{(\mu)}$ and such that $Q^{(p)}$ contains $Q^{(p-1)}$ and is contained in $Q^{(p+1)}$. $Q^{(1)}$ is the prime module determined by the spread of $Q^{(\mu)}$ and is unique; but the intermediate modules $Q^{(2)}, Q^{(3)}, \ldots, Q^{(\mu-1)}$ are to a great extent indeterminate (M, p. 89). Thus $Q^{(1)}, Q^{(2)}, \ldots, Q^{(\mu)}$ may be regarded as successive stages in constructing $Q^{(\mu)}$. *Two primary modules with the same spread and the same multiplicity such that one contains the other must be the same module.*

42. Deductions from Lasker's Theorem. *A module of rank n resolves into simple (primary) modules of which it is the product* (§ 25).

If M' does not contain any relevant spread of M then $M/M' = M$. Let $M/M' = M''$. Then since $M' M''$ contains M, and M' does not contain any relevant spread of M, M'' contains all the relevant primary modules into which M resolves, i.e. $M'' = M$.

It follows that if $M/M' \neq M$, M' must contain a relevant spread of M. *Thus if a polynomial F exists such that $(x_1 - a_1) F, (x_2 - a_2) F, \ldots, (x_n - a_n) F$ are all members of M, while F is not, M contains a relevant simple module whose spread is the point $P(a_1, a_2, \ldots, a_n)$; for $M/P \neq M$.*

Example. The module $M = (x_1^3, x_2^3, \overline{x_1^2 + x_2^2} x_4 + x_1 x_2 x_3)$ has a relevant simple module at the origin; for $x_i x_1^2 x_2^2$ is a member of $M (i = 1, 2, 3, 4)$, but $x_1^2 x_2^2$ is not. The simplest corresponding imbedded primary module, not contained in the L.C.M. of all the other relevant primary modules of M, is (x_1^3, x_2^3, x_3, x_4); cf. Ex. iii, § 17. This example shows that *it is possible for a mixed module M to contain a relevant primary module of higher rank than the number of members in a basis of M.* For the rank of (x_1^3, x_2^3, x_3, x_4) is 4.

If M is an H-module not having a relevant simple module at the origin the variables can be subjected to such a linear homogeneous substitution that x_n will not contain any relevant spread of M, and we shall then have $M/(x_n) = M$, and M will be equivalent to $M_{x_n=1}$ (§ 38). *Thus the only condition (remaining permanent under a linear substitution) that an H-module M may be equivalent to the module $M_{x_n=1}$ is that M should not contain a relevant simple module.*

A simple H-module M is not equivalent to $M_{x_n=1}$; in fact $M_{x_n=1}$ is in this case the module (1).

If M' contains any relevant spread of M then $M/M' \neq M$. Let $M = [Q_1, Q_2, \ldots, Q_k]$, and let M' contain the spread of Q_i. Then some power M'^γ of M' contains Q_i (§ 32), and $Q_i/M'^\gamma = (1)$. Hence the spread of Q_i is not a relevant spread of

$$M/M'^\gamma = [Q_1/M'^\gamma, Q_2/M'^\gamma, \ldots, Q_k/M'^\gamma], \text{ § 28};$$

and consequently $M/M'^\gamma \neq M$. Hence also $M/M' \neq M$; for if $M/M' = M$ then $M/M'^\gamma = M$.

It follows that if $M/M' = M$ then M' does not contain any relevant spread of M. If M_0 is the H-module equivalent to M we know that $M_0/(x_0) = M_0$ (§ 38); hence x_0 does not contain any relevant spread of M_0, i.e. *no module has a relevant spread at infinity.*

If M, M' are any two modules such that M resolves into isolated primary modules only, viz. Q_1, Q_2, ..., Q_k, and (M, M') into primary modules Q_1', Q_2', ..., Q_l', of which Q_1', Q_2', ..., Q_k have the same spreads as Q_1, Q_2, ..., Q_k respectively, then

$$M/M' = [Q_1/Q_1',\ Q_2/Q_2',\ ...,\ Q_k/Q_k'].$$

The spread of (M, M') is contained in the spread of M; and it is to be understood that if (M, M') does not contain the spread of Q_i, then $Q_i' = (1)$. The spreads of Q_{k+1}', ..., Q_l' are contained in those of Q_1, Q_2, ..., Q_k, but do not contain any of the latter. Now we have

$$M/M' = M/(M, M') = [Q_1, Q_2, ..., Q_k]/[Q_1', Q_2', ..., Q_l'].$$

Hence the theorem follows, by the second part of § 28, provided

$$Q_i/[Q_1', Q_2', ..., Q_l'] = Q_i/Q_i'.$$

This is true; for Q_i/Q_i' contains $Q_i/[Q_1', Q_2', ..., Q_l']$, since $[Q_1', Q_2', ..., Q_l']$ contains Q_i', and, for a similar reason, $Q_i/[Q_1', Q_2', ..., Q_l']$ contains $Q_i/Q_1'Q_2'...Q_l'$ or Q_i/Q_i'.

43. *If a module M of rank r is regarded as a module $M^{(s)}$ in s variables x_1, x_2, ..., x_s, while x_{s+1}, ..., x_n are regarded as parameters; and if $F^{(s)}$ is a whole member of $M^{(s)}$, that is, a whole function of the parameters as well as of the variables, then $F^{(s)}$, regarded as a polynomial in x_1, x_2, ..., x_n, contains all the relevant primary modules of M of rank $\leqslant s$; and conversely, any polynomial which contains all these primary modules is a member of $M^{(s)}$. The most important case is that in which $s = r$.*

In other words, to treat a module M as a module in s variables has the sole effect of eliminating all the primary modules of M of rank $> s$; and when $s < r$ it reduces M to the module (1).

Let $M = (F_1, F_2, ..., F_k)$; then $F^{(s)} = A_1F_1 + A_2F_2 + ... + A_kF_k$, where A_1, A_2, ..., A_k are whole functions of x_1, x_2, ..., x_s and rational functions of x_{s+1}, ..., x_n, with a common denominator $D^{(s)}$. Hence $D^{(s)}F^{(s)} = 0 \bmod M$, and $F^{(s)}$ contains all the primary modules of M of rank $\leqslant s$, since $D^{(s)}$ does not contain any of their spreads.

Conversely, if $F^{(s)}$ contains all the primary modules of M of rank $\leqslant s$, and $D^{(s)}$, a whole function of x_{s+1}, ..., x_n only, contains all the primary modules of M of rank $> s$, then $D^{(s)}F^{(s)} = 0 \bmod M$, and $F^{(s)} = 0 \bmod M^{(s)}$, since $D^{(s)}$ in respect to $M^{(s)}$ does not involve the variables.

The module $M^{(r)}$ resolves into simple modules, any primary module of M of rank r and order d contributing d simple modules to $M^{(r)}$. By finding these simple modules we are able to find the primary modules of M of rank r; and this completely resolves M if M is unmixed.

44. *If M is a module of rank $r < n$ and no-one of the modules M, $(M, x_n - a_n)$, $(M, x_{n-1} - a_{n-1}, x_n - a_n)$, ... $(M, x_{r+2} - a_{r+2}, ..., x_n - a_n)$ contains a relevant simple module $(a_{r+2}, ..., a_n$ having non-special values) then M is unmixed. In the contrary case M is mixed.*

This theorem will be used later for proving that certain modules are unmixed. We shall prove first that if M is mixed and does not contain a relevant simple module then $(M, x_n - a_n)$ is mixed. Let M' be the prime module determined by a relevant spread of M of rank $> r$ and $< n$, since M is mixed and has no relevant spread of rank n. To prove that $(M, x_n - a_n)$ is mixed it is sufficient to show that $(M', x_n - a_n)$ contains a relevant spread of $(M, x_n - a_n)$.

Suppose this is not the case; then (§ 42)
$$(M, x_n - a_n)/(M', x_n - a_n) = (M, x_n - a_n),$$
i.e.
$$(M, x_n - a_n)/M' = (M, x_n - a_n),$$
and therefore M/M' contains $(M, x_n - a_n)$. Let F be any member of M/M' and $(F_1, F_2, ..., F_k)$ a basis of M; then
$$F = A_1 F_1 + ... + A_k F_k \bmod (x_n - a_n),$$
i.e.
$$F_{x_n = a_n} = (A_1 F_1 + ... + A_k F_k)_{x_n = a_n}.$$
Here we may regard a_n as a parameter replacing x_n. Hence F is a member of M regarded as a module in $n - 1$ variables, and therefore contains all the primary modules of M of rank $\leqslant n - 1$ (§ 43); i.e. $F = 0 \bmod M$. Hence M/M' contains M, which is not true. It follows that $(M, x_n - a_n)$ is mixed in general, i.e. if a_n has a non-special value. By the same reasoning, if $(M, x_n - a_n)$ does not contain a relevant simple module, $(M, x_{n-1} - a_{n-1}, x_n - a_n)$ is mixed, and so on. Finally if $(M, x_{r+2} - a_{r+2}, ..., x_n - a_n)$ is mixed it must contain a relevant simple module since it is of rank $n - 1$. Hence if M is mixed one of the above modules contains a relevant simple module. It follows that if no-one of the modules contains a relevant simple module, then M is unmixed.

Conversely if one of the above modules contains a relevant simple module (or more generally if one is mixed) then M is mixed. Suppose for instance that $(M, x_n - a_n)$ is mixed. Then since $(M, x_n - a_n)$ is of rank $r + 1$ it has a relevant spread of rank $\geqslant r + 2$. Hence there is a whole function ϕ of $x_{r+1}, ..., x_{n-1}$ only containing this spread, and a polynomial F in $x_1, x_2, ..., x_n$ such that
$$\phi F = 0 \bmod (M, x_n - a_n), \text{ while } F \neq 0 \bmod (M, x_n - a_n).$$

Let $(F_1, F_2, ..., F_k)$ be a basis of M. Then

$$\phi F = A_1 F_1 + ... + A_k F_k \bmod (x_n - a_n),$$

where we may assume that F, ϕ, A_1, ..., A_k are whole functions of a_n as well as of x_1, x_2, ... x_n. Putting $x_n = a_n$,

$$(\phi F)_{x_n = a_n} = (A_1 F_1 + ... + A_k F_k)_{x_n = a_n}.$$

In this we can replace a_n by x_n, and we then have

$$(\phi F)_{a_n = x_n} = 0 \bmod M.$$

But $\phi_{a_n = x_n}$ is a whole function of $x_{r+1}, ..., x_n$ only, and $F_{a_n = x_n} \neq 0 \bmod M$, since $F \neq 0 \bmod (M, x_n - a_n)$. Hence M is mixed. This completes the proof of the theorem.

If M is unmixed all the modules are unmixed; nevertheless *if $a_{r+2}, ..., a_n$ have special values, some of the modules may be mixed notwithstanding that M is unmixed.*

Example. The module

$$M = (u_0 u_4 - u_1 u_3,\ u_1^3 - u_0^2 u_3,\ u_3^3 - u_1 u_4^2,\ u_1^2 u_4 - u_0 u_3^2)$$

is prime and of rank 2 $\left(\text{its spread being given by } \dfrac{u_0}{\lambda^0} = \dfrac{u_1}{\lambda^1} = \dfrac{u_3}{\lambda^3} = \dfrac{u_4}{\lambda^4}\right)$ while the module $(M, c_4 u_0 + c_3 u_1 + c_1 u_3 + c_0 u_4)$ is of rank 3 and mixed. For the latter has $u_0 F$, $u_1 F$, $u_3 F$, $u_4 F$ as members, where

$$F = c_4 u_1^2 + c_3 u_0 u_3 + c_1 u_1 u_4 + c_0 u_3^2 \neq 0 \bmod (M, c_4 u_0 + c_3 u_1 + c_1 u_3 + c_0 u_4).$$

Hence if u_0, u_1, u_3, u_4 are linear functions of x_1, x_2, x_3, x_4 and (a_1, a_2, a_3, a_4) their common point, the module $(M, x_4 - a_4)$ is mixed notwithstanding that M is unmixed (cf. § 89, end).

45. *If M contains a relevant simple module at the point $(a_1, a_2, ..., a_n)$ then $(M, x_n - a_n)$ contains a relevant simple module at the same point.*

Let u, u', u'', ... be linear functions of x_1, x_2, ..., x_n containing the point $(a_1, a_2, ..., a_n)$ and no other relevant spread of M. Suppose that $(M, x_n - a_n)$ does not contain a relevant simple module at $(a_1, a_2, ..., a_n)$; then it may be assumed that (M, u), (M, u'), (M, u''), ... do not either. Let F be a polynomial such that $uF = 0 \bmod M$ and $F \neq 0 \bmod M$.

Then $\qquad\qquad uF = 0 \bmod (M, u'),$

therefore $\qquad F = 0 \bmod (M, u') = u'F'' \bmod M,$

therefore $\qquad uu'F'' = 0 \bmod M = 0 \bmod (M, u''),$

therefore $\qquad F'' = 0 \bmod (M, u'') = u''F''' \bmod M,$

and $\qquad\qquad F = u'u''F''' \bmod M.$

Similarly $\qquad F = u'u''...u^{(t)}F^{(t)} \bmod M.$

Now l can be chosen so great that $u'u''...u^{(l)}$ contains the relevant simple module of M at $(a_1, a_2, ..., a_n)$; and since F contains all the other relevant primary modules of M we have $F = 0 \bmod M$, which is not true. Hence $(M, x_n - a_n)$ does contain a relevant simple module at the point $(a_1, a_2, ..., a_n)$.

46. The Hilbert-Netto Theorem (H₁, Ne). *If M' is any module containing the spread of a given module M some finite power of M' contains M.*

For M' contains all the relevant spreads of M and some finite power of M' contains all the relevant primary modules of M (§ 32) and therefore contains M.

The theorem is proved in (Ne) for the case of two variables and in (H₁) for the general case.

47. *Definition.* A module of rank r having a basis consisting of r members only is called a *module of the principal class* (Kr, p. 80). Hence a module $(F_1, F_2, ..., F_r)$ of rank r is of the principal class.

It is possible for the resultant of a module of the principal class to vanish identically. An example is given at the end of § 12.

The H-module equivalent to a given module of the principal class is not necessarily of the principal class, e.g. the H-module equivalent to $(x_1^2, x_2 + x_1 x_3)$ has four members in its basis $(x_1^2, x_1 x_2, x_2^2, x_2 x_0 + x_1 x_3)$, § 38.

A proper module is of rank $\leqslant n$ and $\geqslant 1$.

A proper module with a basis consisting of r members is of rank $\leqslant r$ (cf. ex. § 42); for the module contains some point P in the finite region and a spread of dimensions $n - r$ at least through any such point. Nevertheless a module with a basis of two or more members may be the non-proper module (1); e.g. $(F, 1 + F) = (1)$.

The unit module is sometimes said to be of rank $n + 1$; but it is better to say that it is without rank, and that no module is of rank $> n$. In the absolute theory a module can be of rank $n + 1$.

If $(F_1, F_2, ..., F_r)$ is of rank r it does not necessarily follow that $(F_2, F_3, ..., F_r)$ is of rank $r - 1$. Thus $(f, f_1 + ff_1, f_2 + ff_2)$ is the same as (f, f_1, f_2), and can be of rank 3, while $(f_1 + ff_1, f_2 + ff_2)$ contains $(1 + f)$ and is of rank 1. If however the series $F_1, F_2, ..., F_r$ is suitably modified beforehand (§ 37) then $(F_{s+1}, ..., F_r)$ will be of rank $r - s$ if $(F_1, F_2, ..., F_r)$ is of rank r. It will be sufficient to prove that $(F_2 + a_2 F_1, F_3 + a_3 F_1, ..., F_r + a_r F_1)$ is of rank $r - 1$ when $a_2, a_3, ..., a_r$

are (at first) undetermined constants. If it is of rank $s < r-1$ then the module $(\phi_1, \phi_2, ..., \phi_s)$ is of rank s, where

$$\phi_i = \lambda_{i2}(F_2 + a_2 F_1) + \lambda_{i3}(F_3 + a_3 F_1) + ... + \lambda_{ir}(F_r + a_r F_1)$$
$$= \lambda_{i1} F_1 + \lambda_{i2} F_2 + ... + \lambda_{ir} F_r$$
$$(\lambda_{i1} = \lambda_{i2} a_2 + \lambda_{i3} a_3 + ... + \lambda_{ir} a_r, \quad i = 1, 2, ..., s),$$

and the λ_{ij} are all arbitrary constants. We may regard the s relations $\lambda_{i1} = \lambda_{i2} a_2 + ... + \lambda_{ir} a_r$ as determining the s constants $a_2, a_3, ..., a_{s+1}$, leaving at least a_r $(s + 1 \leqslant r - 1)$ quite arbitrary, whatever the values of the λ_{ij} are. Now some spread of $(\phi_1, \phi_2, ..., \phi_s)$ of rank s is a spread of $(F_2 + a_2 F_1, ..., F_r + a_r F_1)$ and is contained in $F_r + a_r F_1$, and therefore in F_1 (since a_r is independent of the λ_{ij}), and in each of $F_2, F_3, ..., F_r$. This would make $(F_1, F_2, ..., F_r)$ of rank s, which is not the case.

Unmixed Modules

48. A useful test as to whether a given module is mixed or unmixed is proved in § 44.

Theorem. *A module of the principal class is unmixed.* Lasker proves this for H-modules (L, p. 58). The following is a general proof.

It is clear that any module of rank n is unmixed, since it resolves into primary modules which are all of rank n. Also a module of the principal class of rank 1 is unmixed. Hence the theorem is true for two variables, since in this case the module can only be of rank 1 or 2. We shall assume the theorem true for $n - 1$ variables and prove it for n variables. We also assume that the members of the basis have been modified if necessary so that, when $(F_1, F_2, ..., F_r)$ is of rank r, $(F_2, F_3, ..., F_r)$ is of rank $r - 1$ (§ 47).

We prove first that a module $M = (F_1, F_2, ..., F_r)$ of rank $r < n$ cannot contain any relevant simple module by showing that $(x_n - c_n) F = 0 \bmod M$ requires $F = 0 \bmod M$ no matter what value, special or otherwise, c_n may have.

Let $(x_n - c_n) F = X_1 F_1 + X_2 F_2 + ... + X_r F_r$;

then $(X_1 F_1 + X_2 F_2 + ... + X_r F_r)_{x_n = c_n} = 0,$

and $(X_1 F_1)_{x_n = c_n} = 0 \bmod (F_2, F_3, ..., F_r)_{x_n = c_n}.$

But $(F_2, F_3, ..., F_r)_{x_n = c_n}$ is a module of rank $r - 1$ in $n - 1$ variables, so that (by the assumption) all its relevant spreads are of rank $r - 1$, and $(F_1)_{x_n = c_n}$ does not contain any of them. Hence

$$(X_1)_{x_n = c_n} = 0 \bmod (F_2, F_3, ..., F_r)_{x_n = c_n},$$

i.e. $X_1 = X_{12} F_2 + X_{13} F_3 + ... + X_{1r} F_r + (x_n - c_n) Y_1.$

Substituting this value for X_1 in the equation

$$(X_1 F_1 + X_2 F_2 + \ldots + X_r F_r)_{x_n = c_n} = 0,$$

we have $\{(X_2 + X_{12} F_1) F_2 + \ldots + (X_r + X_{1r} F_1) F_r\}_{x_n = c_n} = 0.$

Hence, by the same reasoning as before,

$$X_2 + X_{12} F_1 = X_{23} F_3 + \ldots + X_{2r} F_r + (x_n - c_n) Y_2,$$

$$X_3 + X_{13} F_1 + X_{23} F_2 = X_{34} F_4 + \ldots + X_{3r} F_r + (x_n - c_n) Y_3,$$

$$\cdots\cdots\cdots\cdots\cdots\cdots\cdots\cdots\cdots\cdots\cdots$$

$$X_r + X_{1r} F_1 + X_{2r} F_2 + \ldots + X_{r-1, r} F_{r-1} = (x_n - c_n) Y_r.$$

Multiplying these equations by F_1, F_2, ..., F_r and adding we have

$$X_1 F_1 + X_2 F_2 + \ldots + X_r F_r = (x_n - c_n) (Y_1 F_1 + Y_2 F_2 + \ldots + Y_r F_r),$$

all the terms $\Sigma X_{ij} F_i F_j \, (i < j)$ cancelling from both sides. It follows that

$$F = Y_1 F_1 + Y_2 F_2 + \ldots + Y_r F_r = 0 \bmod M,$$

and that $(F_1, F_2, \ldots, F_r)$ does not contain any relevant simple module.

Now if $(F_1, F_2, \ldots, F_r)$ were mixed then for some value of $s \geqslant r + 2$ the module $(F_1, \ldots, F_r, x_s - a_s, \ldots, x_n - a_n)$ would contain a relevant simple module (§ 44); but it does not, because it is of the principal class. Hence $(F_1, F_2, \ldots, F_r)$ is unmixed.

49. Deductions from the theorem.   *A basis $(F_1, F_2, \ldots, F_r)$ of a module M of the principal class of rank r is an H-basis of M or not, and an H-basis of $M^{(r)}$ or not, according as the H-module determined by the terms of highest degree in $F_1, F_2, \ldots, F_r$ is of rank r or not.*

Let M_0 be the H-module in $x_1, x_2, \ldots, x_n, x_0$ corresponding to the basis $(F_1, F_2, \ldots, F_r)$, so that $(M_0)_{x_0 = 0}$ is the H-module mentioned in the enunciation. Let $(M_0)_{x_0 = 0}$ be of rank r. Then it follows by the same reasoning as in the theorem that $x_0 F_0 = 0 \bmod M_0$ requires $F_0 = 0 \bmod M_0$. Hence M_0 is equivalent to M (§ 38), i.e. $(F_1, F_2, \ldots, F_r)$ is an H-basis of M. It is also an H-basis of $M^{(r)}$. This follows in the same way by considering the H-module $M_0^{(r)}$ in $x_1, x_2, \ldots, x_r, x_0$ corresponding to $(F_1, F_2, \ldots, F_r)$ regarded as a basis of $M^{(r)}$. The module $(M_0^{(r)})_{x_0 = 0}$ is a simple H-module not involving $x_{r+1}, \ldots, x_n$.

If on the contrary $(M_0)_{x_0 = 0}$ is not of rank r it is of rank $< r$, and x_0 contains a relevant spread of M_0 of rank $\leqslant r$, so that $M_0/(x_0) \neq M_0$ and M_0 is not equivalent to M (§ 38). Hence $(F_1, F_2, \ldots, F_r)$ is not an H-basis of M or of $M^{(r)}$.

If $(F_1, F_2, \ldots, F_k)$ is an H-basis of a module of rank r the H-module determined by the terms of highest degree in $F_1, F_2, \ldots, F_k$ is of rank r.

But the converse is not true in general when $k > r$; i.e. if the module determined by the terms of highest degree in F_1, F_2, ..., F_k is of the same rank r as the module $(F_1, F_2, ..., F_k)$ the basis $(F_1, F_2, ..., F_k)$ is not in general an H-basis when $k > r$.

50. *Any power of a module of the principal class is unmixed.*

Let the module be $M = (F_1, F_2, ..., F_r)$ of rank r. The spread of M^γ is the same as the spread of M. Hence it will be sufficient to show that $AF = 0 \bmod M^\gamma$ requires $F = 0 \bmod M^\gamma$ provided A does not contain any relevant spread of M. When $\gamma = 2$ we have

$$AF = 0 \bmod M^2; \text{ hence } F = 0 \bmod M = A_1 F_1 + ... + A_r F_r,$$

and $\quad A\,(A_1 F_1 + ... + A_r F_r) = 0 \bmod M^2 = F_1 F^{(1)} \bmod (F_2, ..., F_r)^2,$

where $\qquad\qquad\qquad F^{(1)} = 0 \bmod M.$

Hence $\qquad\quad (AA_1 - F^{(1)})\, F_1 = 0 \bmod (F_2, ..., F_r),$

$$AA_1 - F^{(1)} = 0 \bmod (F_2, ..., F_r),$$

$$AA_1 = 0 \bmod M, \text{ and } A_1 = 0 \bmod M.$$

Similarly $A_i = 0 \bmod M$, and $F = A_1 F_1 + ... + A_r F_r = 0 \bmod M^2$.

Next suppose $\gamma = 3$. Then since

$$AF = 0 \bmod M^3,$$

$$F = 0 \bmod M^2 = F_1 F^{(1)} + \phi^{(2)},$$

where $\quad F^{(1)} = A_1 F_1 + ... + A_r F_r$, and $\phi^{(2)} = 0 \bmod (F_2, ..., F_r)^2$.

Now $\quad A\,(F_1 F^{(1)} + \phi^{(2)}) = 0 \bmod M^3 = F_1 F^{(2)} \bmod (F_2, ..., F_r)^3,$

where $\qquad\qquad\qquad F^{(2)} = 0 \bmod M^2;$

hence $\qquad\quad (AF^{(1)} - F^{(2)})\, F_1 = 0 \bmod (F_2, ..., F_r)^2,$

$$AF^{(1)} - F^{(2)} = 0 \bmod (F_2, ..., F_r)^2,$$

$$F^{(1)} = 0 \bmod M^2.$$

Thus every coefficient A_i in $F^{(1)} (= A_1 F_1 + ... + A_r F_r)$ is a member of M (as proved when $\gamma = 2$), i.e. every coefficient of the terms of $F = F_1 F^{(1)} + \phi^{(2)}$ furnished by $F_1 F^{(1)}$ is a member of M; and the same must therefore be true of the terms of F furnished by $\phi^{(2)}$. Hence

$$F = 0 \bmod M^3.$$

Similarly, if $AF = 0 \bmod M^\gamma$, and the theorem is assumed true for $M^{\gamma-1}$ we have $F = 0 \bmod M^{\gamma-1} = F_1 F^{(\gamma-2)} + \phi^{(\gamma-1)}$, and can prove that every coefficient in $F^{(\gamma-2)}$ and $\phi^{(\gamma-1)}$ is a member of M. Hence

$$F = 0 \bmod M^\gamma.$$

51. *If M is a module of the principal class which resolves into prime modules the module whose members consist of all polynomials having a γ-point at every point of M is the module M^γ.*

The theorem is true when $\gamma = 1$. We shall prove it for M^γ assuming it for $M^{\gamma-1}$. Let $M = (F_1, F_2, \ldots, F_r)$ be of rank r and let $F^{(\gamma)}$ be any polynomial with a γ-point at every point of M.

Then
$$F^{(\gamma)} = 0 \bmod M^{\gamma-1},$$

i.e. $F^{(\gamma)} = \Sigma A_{p_1, p_2, \ldots, p_r} F_1^{p_1} F_2^{p_2} \ldots F_r^{p_r}$, where $p_1 + p_2 + \ldots + p_r = \gamma - 1$. Take $\xi_1, \xi_2, \ldots, \xi_n$ for the variables instead of $x_1, x_2, \ldots, x_n$, and move the origin to any point $(x_1, x_2, \ldots, x_n)$ of M. Then F_1 becomes

$$F_1(\xi_1 + x_1, \ldots, \xi_n + x_n) = \xi_1 \frac{\partial F_1}{\partial x_1} + \ldots + \xi_n \frac{\partial F_1}{\partial x_n} + \tfrac{1}{2}\xi_1^2 \frac{\partial^2 F_1}{\partial x_1^2} + \ldots,$$

and the terms of lowest degree in $F^{(\gamma)}$ are

$$\Sigma A_{p_1, p_2, \ldots, p_r} \left(\xi_1 \frac{\partial F_1}{\partial x_1} + \ldots + \xi_n \frac{\partial F_1}{\partial x_n} \right)^{p_1} \ldots \left(\xi_1 \frac{\partial F_r}{\partial x_1} + \ldots + \xi_n \frac{\partial F_r}{\partial x_n} \right)^{p_r},$$

where $A_{p_1, p_2, \ldots, p_r}$ have their original values as functions of $x_1, x_2, \ldots, x_n$. This last expression is of degree $\gamma - 1$ in $\xi_1, \xi_2, \ldots, \xi_n$ and must vanish identically, since $F^{(\gamma)}$ has a γ-point at every point of M. Now the r quantities $\xi_1 \frac{\partial F_i}{\partial x_1} + \ldots + \xi_n \frac{\partial F_i}{\partial x_n}$ $(i = 1, 2, \ldots, r)$ are either capable of taking any r values ($\xi_1, \ldots, \xi_n$ being undetermined quantities and $x_1, \ldots, x_n$ fixed quantities) or they are not. If they are, every $A_{p_1, p_2, \ldots, p_r}$ vanishes. If they are not, every determinant of the matrix

$$\begin{vmatrix} \dfrac{\partial F_1}{\partial x_1} & \dfrac{\partial F_1}{\partial x_2} & \cdots & \dfrac{\partial F_1}{\partial x_n} \\ \cdots\cdots\cdots\cdots \\ \dfrac{\partial F_r}{\partial x_1} & \dfrac{\partial F_r}{\partial x_2} & \cdots & \dfrac{\partial F_r}{\partial x_n} \end{vmatrix}$$

vanishes, i.e. $(x_1, x_2, \ldots, x_n)$ is a singular point of M (§ 29). Hence every $A_{p_1, p_2, \ldots, p_r}$ vanishes for every non-singular point of M and is therefore a member of M (§ 22). Hence $F^{(\gamma)} = 0 \bmod M^\gamma$, which proves the theorem.

52. *Definition.* The module whose basis consists of all the determinants of the matrix

$$\begin{vmatrix} u_1, & u_2, & \ldots, & u_k \\ v_1, & v_2, & \ldots, & v_k \\ \cdots\cdots\cdots\cdots \end{vmatrix},$$

where the elements u, v, w,... are polynomials, will be denoted by

$$\begin{pmatrix} u_1, & u_2, & ..., & u_k \\ v_1, & v_2, & ..., & v_k \\ \multicolumn{4}{c}{\cdots\cdots\cdots\cdots} \end{pmatrix}.$$

This is only an extension of the notation $(F_1, F_2, ..., F_k)$ for a module M.

If M_1 is a prime module of rank r, and $F_1, F_2, ..., F_r$ any r members of M_1 such that $M = (F_1, F_2, ..., F_r)$ resolves into M_1 and a second prime module M_1' of rank r, then it may happen that M_1' must have a certain fixed spread in common with M_1 irrespective of the choice of $F_1, F_2, ..., F_r$. Such a spread (if any exists) must be a singular spread of M_1; but it does not necessarily follow from M_1 having a singular spread that M_1' must contain the spread; it depends on the nature of the singularity. *If M_1' does not cut M_1 in a fixed spread then M_1^γ is unmixed, and is the module whose members consist of all polynomials having a γ-point at every point of M_1. In the contrary case some power M_1^γ of M_1 will be mixed and will have the fixed spread in which M_1' cuts it as a relevant imbedded spread, while polynomials $F^{(\gamma)}$ having a γ-point at every point of M_1, but not members of M_1^γ, will exist.*

Example i.   The square of the prime module M_1 determined by an irreducible curve in space of three dimensions having a triple* point, the tangents at which do not lie in one plane, is mixed; and there is consequently a surface having a 2-point at every point of the curve which is not a member of M_1^2.

Thus if

$$M_1 = \begin{pmatrix} x_1, & x_2, & x_3 \\ x_2, & x_3, & x_1^2 \end{pmatrix} = (x_1 x_3 - x_2^2, \; x_2 x_3 - x_1^3, \; x_3^2 - x_1^2 x_2),$$

the surface $(x_2 x_3 - x_1^3)^2 - (x_2^2 - x_1 x_3)(x_3^2 - x_1^2 x_2)$, after removal of the factor x_1, will have a 2-point at every point of M_1, but is not a member of M_1^2; for the surface has only a 3-point at the origin, whereas every member of M_1^2 has a 4-point.

Example ii.   If $M_1 = \begin{pmatrix} u_1, & u_2, & u_3, & u_4 \\ v_1, & v_2, & v_3, & v_4 \\ w_1, & w_2, & w_3. & w_4 \end{pmatrix} = (F_1, F_2, F_3, F_4),$

* A triple point is not a 3-point. The general member of M_1 has only a 2-point at the triple point of the curve.

It is evident that the module whose members consist of all polynomials having a γ-point at every point of a given irreducible spread is primary and unmixed.

where each u, v, w is a homogeneous linear function of x_1, x_2, x_3, x_4, M_1 being a prime module of rank 2, we have

$$u_1 F_1 + u_2 F_2 + u_3 F_3 + u_4 F_4 = 0,$$

and two other similar identities. From these we can find the continued ratio $x_1 : x_2 : x_3 : x_4$ as the ratio of four members of M_1^3 by expressing each u, v, w in full. The common factor of these four members is a polynomial of degree 8 having a 3-point at every point of M_1, but not a member of M_1^3. In this example M_1^3 is mixed while M_1^2 is unmixed.

53. Theorem. *The module with a basis of r rows and k columns*

$$M = \begin{pmatrix} u_1, & u_2, & \dots, & u_k \\ v_1, & v_2, & \dots, & v_k \\ w_1, & w_2, & \dots, & w_k \\ \dots\dots\dots\dots \end{pmatrix}$$

is of rank $\leqslant k - r + 1$ $(0 < k - r + 1 \leqslant n)$, and if of rank $k - r + 1$ is unmixed.

Also if $D_{p_1, p_2, \dots, p_r}$ denotes the determinant formed by the p_1^{th}, p_2^{th}, ..., p_r^{th} columns of the basis, the general solution of the equation

$$\Sigma D_{p_1, p_2, \dots, p_r} X_{p_1, p_2, \dots, p_r} = 0 \quad (p_1, p_2, \dots, p_r = 1, 2, \dots, k)$$

is
$$X_{p_1, p_2, \dots, p_r} = \sum_{p=1}^{p=k} U_{p_1, \dots, p_r, p} \, u_p + \sum_{p=1}^{p=k} V_{p_1, p_2, \dots, p_r, p} \, v_p + \dots,$$

where $U_{p_1, \dots, p_r, p}, V_{p_1, \dots, p_r, p}, \dots$ are arbitrary polynomials subject with the unknowns $X_{p_1, p_2, \dots, p_r}$ to the same law of signs as the determinants $D_{p_1, p_2, \dots, p_r}$, viz. each $X_{p_1, p_2, \dots, p_r}, U_{p_1, p_2, \dots, p_r, p}, \dots$ changes in sign (but not in magnitude) for each interchange of any pair of suffixes $p_1, \dots, p_r, p$.

These two theorems will be proved together by a double process of induction. Assuming both theorems for $r - 1$ rows and $k - 1$ columns, and also for r rows and $k - 1$ columns, we prove both theorems for r rows and k columns. Both theorems have been proved for $r = 1$ in § 48.

It is understood that M is a proper module, i.e. the determinants of its basis all vanish for some point whose coordinates are finite, but do not all vanish identically. After proving that M is of rank $\leqslant k - r + 1$ we assume that if M is of rank $k - r + 1$ the module.

$$\begin{pmatrix} u_2 + a_2 u_1, & u_3 + a_3 u_1, & \dots, & u_k + a_k u_1 \\ v_2 + a_2 v_1, & v_3 + a_3 v_1, & \dots, & v_k + a_k v_1 \\ \dots\dots\dots\dots\dots\dots\dots\dots\dots\dots\dots\dots\dots \end{pmatrix},$$

where a_2, a_3, ..., a_k are suitably chosen constants or polynomials, is of rank $k - r$. This can be proved in a similar way to the corresponding property in § 47. We shall also suppose the matrix to have been so modified beforehand that if the first $s \leqslant k - r$ columns are removed the rank diminishes by s. It can be shown that the second part of the theorem is true before modification if it is true after. The same is true of the first part of the theorem, since the modification of the basis does not alter the module.

The general proof will be sufficiently indicated if we suppose M to have 3 rows and 5 columns. Then

$$M = \begin{pmatrix} u_1 & u_2 & u_3 & u_4 & u_5 \\ v_1 & v_2 & v_3 & v_4 & v_5 \\ w_1 & w_2 & w_3 & w_4 & w_5 \end{pmatrix}$$

and we assume both parts of the theorem for the module

$$M_1 = \begin{pmatrix} u_2 & u_3 & u_4 & u_5 \\ v_2 & v_3 & v_4 & v_5 \\ w_2 & w_3 & w_4 & w_5 \end{pmatrix}$$

and also for $$M_1' = \begin{pmatrix} u_2 & u_3 & u_4 & u_5 \\ v_2 & v_3 & v_4 & v_5 \end{pmatrix}.$$

If A, B, C are the determinants of the matrix formed by the last two columns of the basis of M, we have

$$A u_i + B v_i + C w_i = D_{i45} \quad (i = 1, 2, 3, 4, 5).$$

Giving to i the values p_1, p_2, p_3 and solving for C (or D_{45}) we have

$$D_{45} D_{p_1 p_2 p_3} = D_{p_2 p_3} D_{p_1 45} + D_{p_3 p_1} D_{p_2 45} + D_{p_1 p_2} D_{p_3 45},$$

where $D_{p_1 p_2}$ denotes the determinant $\begin{vmatrix} u_{p_1} & u_{p_2} \\ v_{p_1} & v_{p_2} \end{vmatrix}$. This shows that every determinant $D_{p_1 p_2 p_3}$ when multiplied by D_{45} is of the form $X_1 D_{145} + X_2 D_{245} + X_3 D_{345}$. Hence if there is a point of the module M for which D_{45} does not vanish the module must have a spread of rank $\leqslant 3$ (or $k - r + 1$) through that point. If however D_{45} contains the whole of the spread of M we move the origin to a point of the spread and modify the last row of the basis by the other rows so as to make the constant terms in the elements of the last row all zero. After doing this we change u_4, u_5, v_4, v_5 in the first two rows only to $u_4 + a$, $u_5 + b$, $v_4 + c$, $v_5 + d$, where a, b, c, d are constants. We thus get a new module containing the origin such that the new D_{45} does not contain the origin. This new module has a spread of rank $\leqslant 3$ through the origin; and since this is true for general values of a, b, c, d, it is still true when we put $a = b = c = d = 0$; for no diminution in the dimensions of the spread through the origin, i.e. no increase in the rank, could be produced by giving special values to a, b, c, d. Hence M is of rank $\leqslant 3$; and we have to prove that M is unmixed if its rank is 3.

Consider the equation $\Sigma D_{p_1 p_2 p_3} X_{p_1 p_2 p_3} = 0$ in which we suppose $p_1 < p_2 < p_3$, so that each term occurs once and once only. Multiplying by D_{45} we have

$$\Sigma (D_{p_2 p_3} D_{p_1 45} + D_{p_3 p_1} D_{p_2 45} + D_{p_1 p_2} D_{p_3 45}) X_{p_1 p_2 p_3} = 0.$$

In this the terms containing D_{145} are obtained by putting $p_1 = 1$ and giving p_2, p_3 the values $(2, 3), (2, 4), (2, 5), (3, 4), (3, 5), (4, 5)$, viz. $D_{145} (D_{23} X_{123} + D_{24} X_{124} + D_{25} X_{125} + D_{34} X_{134} + D_{35} X_{135} + D_{45} X_{145})$, and this is a member of (D_{245}, D_{345}) and therefore of M_1. But M_1 is unmixed and of rank 2, and D_{145} does not contain any of its relevant spreads; for if, after modification of the last two columns of M_1 by the first two, D_{145} contains a relevant spread of M_1 then every $D_{1 p_2 p_3}$ contains the same spread, and consequently M contains the spread and is of rank 2, which is contrary to the data. Hence

$$\Sigma D_{p_2 p_3} X_{1 p_2 p_3} = 0 \bmod M_1$$
$$= D_{234} W'_{234} + D_{235} W'_{235} + D_{245} W'_{245} + D_{345} W'_{345}$$
$$= (D_{34} w_2 + D_{42} w_3 + D_{23} w_4) W'_{234} + \ldots$$
$$= D_{23} (w_4 W'_{234} + w_5 W'_{235}) + \ldots$$
$$= \Sigma D_{p_2 p_3} (W'_{p_2 p_3 2} w_2 + W'_{p_2 p_3 3} w_3 + W'_{p_2 p_3 4} w_4 + W'_{p_2 p_3 5} w_5);$$

or $\Sigma D_{p_2 p_3} X_{p_2 p_3} = 0 \quad (p_2 < p_3 = 2, 3, 4, 5),$

where $X_{p_2 p_3} = X_{1 p_2 p_3} - \underset{p}{\Sigma} W'_{p_2 p_3 p} w_p \quad (p = 2, 3, 4, 5).$

The equation $\Sigma D_{p_2 p_3} X_{p_2 p_3} = 0$ stands in the same relation to M_1' as $\Sigma D_{p_1 p_2 p_3} X_{p_1 p_2 p_3} = 0$ to M, and the general solution is

$$X_{p_2 p_3} = \underset{p}{\Sigma} U'_{p_2 p_3 p} u_p + \underset{p}{\Sigma} V'_{p_2 p_3 p} v_p \quad (p = 2, 3, 4, 5)$$

which gives

$$X_{1 p_2 p_3} = \underset{p}{\Sigma} U'_{p_2 p_3 p} u_p + \underset{p}{\Sigma} V'_{p_2 p_3 p} v_p + \underset{p}{\Sigma} W'_{p_2 p_3 p} w_p \quad (p = 2, 3, 4, 5).$$

Substituting these values for $X_{1 p_2 p_3}$ in the equation

$$\Sigma D_{p_1 p_2 p_3} X_{p_1 p_2 p_3} = 0$$

it becomes, after simplifying,

$$D_{234} (X_{234} + U'_{234} u_1 + V'_{234} v_1 + W'_{234} w_1) + \ldots = 0,$$

an equation in reference to M_1 of which the solution is

$$X_{234} = - U'_{234} u_1 - V'_{234} v_1 - W'_{234} w_1 + \underset{p}{\Sigma} U_{234 p} u_p + \ldots + \ldots \quad (p = 2, 3, 4, 5)$$

and similar expressions for $X_{235}, X_{245}, X_{345}$. If in these and the expressions found for $X_{1 p_2 p_3}$ we put

$$- U'_{p_2 p_3 p} = U_{p_2 p_3 p 1}, \quad - V'_{p_2 p_3 p} = V_{p_2 p_3 p 1}, \quad - W'_{p_2 p_3 p} = W_{p_2 p_3 p 1},$$

we have, for all values of $p_1, p_2, p_3 = 1, 2, 3, 4, 5$,

$$X_{p_1 p_2 p_3} = \sum_p U_{p_1 p_2 p_3 p}\, u_p + \sum_p V_{p_1 p_2 p_3 p}\, v_p + \sum_p W_{p_1 p_2 p_3 p}\, w_p \quad (p = 1, 2, 3, 4, 5),$$

which proves the second part of the theorem for M.

To prove the first part, that M is unmixed, it has to be shown that neither M nor $(M, x_s - a_s, \ldots, x_n - a_n)$ can contain a relevant simple module, where s is any number $\geq k - r + 3$ (§ 44). Let

$$(x_1 - c_1)\, F = 0 \bmod (M, x_s - a_s, \ldots, x_n - a_n)$$

$$= \Sigma D_{p_1, p_2, \ldots, p_r}\, X_{p_1, p_2, \ldots, p_r} \bmod (x_s - a_s, \ldots, x_n - a_n).$$

Then $\quad (\Sigma D_{p_1, p_2, \ldots, p_r}\, X_{p_1, p_2, \ldots, p_r})_{x_1 - c_1 = x_s - a_s = \ldots = x_n - a_n = 0} = 0.$

In putting $x_1 - c_1 = x_s - a_s = \ldots = x_n - a_n = 0$ in M the number of variables is diminished but the rank remains equal to $k - r + 1$. Hence

$$(X_{p_1, p_2, \ldots, p_r})_{x_1 - c_1 = \ldots = 0} = (\sum_p U_{p_1, \ldots, p_r, p}\, u_p + \ldots)_{x_1 - c_1 = \ldots = 0} ;$$

therefore

$$X_{p_1, p_2, \ldots, p_r}$$

$$= \sum_p U_{p_1, \ldots, p_r, p}\, u_p + \ldots + (x_1 - c_1)\, Y_{p_1, p_2, \ldots, p_r} \bmod (x_s - a_s, \ldots, x_n - a_n),$$

and

$$(x_1 - c_1)\, F = (x_1 - c_1)\, \Sigma D_{p_1, p_2, \ldots, p_r}\, Y_{p_1, p_2, \ldots, p_r} \bmod (x_s - a_s, \ldots, x_n - a_n).$$

Hence, since $(x_s - a_s, \ldots, x_n - a_n)$ is a module of the principal class, and $x_1 - c_1$ does not contain its spread,

$$F - \Sigma D_{p_1, p_2, \ldots, p_r}\, Y_{p_1, p_2, \ldots, p_r} = 0 \bmod (x_s - a_s, \ldots, x_n - a_n),$$

and $\qquad\qquad F = 0 \bmod (M, x_s - a_s, \ldots, x_n - a_n).$

Hence $(M, x_s - a_s, \ldots, x_n - a_n)$ cannot contain any relevant simple module, which proves the theorem.

Solution of Homogeneous Linear Equations

54. *Homogeneous linear equations with constants for coefficients.*
In a system of r independent equations with constant coefficients for k
unknowns $X_1, X_2, \ldots, X_k$ there are r' independent solutions, where
$r + r' = k$, and the general solution is expressible in terms of the r'
solutions. The array of the coefficients of the r equations and the
array of the r' solutions together form a square array

$$
\begin{array}{|c|}
\hline
a_{11} \quad a_{12} \ldots a_{1k} \\
\cdots\cdots\cdots\cdots \\
a_{r1} \quad a_{r2} \ldots a_{rk} \\
\hline
b_{11} \quad b_{12} \ldots b_{1k} \\
\cdots\cdots\cdots\cdots \\
b_{r'1} \quad b_{r'2} \ldots b_{r'k} \\
\hline
\end{array}
$$

and the general solution is $X_i = \mu_1 b_{1i} + \mu_2 b_{2i} + \ldots + \mu_{r'} b_{r'i}$ $(i = 1, 2, \ldots, k)$,
where $\mu_1, \mu_2, \ldots, \mu_{r'}$ are arbitrary quantities.

The two arrays are called conjugate arrays; but we shall find it
more convenient to call them *inverse* arrays. Their principal properties
are :—(i) the sum of the products of the elements in any row of one
array with the elements in any row of the other array is zero; (ii) the
determinants of one array are proportional to the complementary determinants of the other array with a rule as regards sign; (iii) the
determinant of the combined arrays is not zero if the elements are
real. We shall not have occasion to use either (ii) or (iii) explicitly.

Homogeneous linear equations with polynomials as coefficients
(H, p. 483). Let there be r independent equations, viz.

$$u_1 X_1 + u_2 X_2 + \ldots + u_k X_k = 0,$$

$$v_1 X_1 + v_2 X_2 + \ldots + v_k X_k = 0, \text{ etc.}$$

Then there is an array of solutions

$$
\begin{array}{|c|}
\hline
f_{11}, f_{12}, \ldots, f_{1k} \\
f_{21}, f_{22}, \ldots, f_{2k} \\
\cdots\cdots\cdots\cdots\cdots \\
f_{l1}, f_{l2}, \ldots, f_{lk} \\
\hline
\end{array}
$$

whose elements are polynomials, such that the general solution is

$$X_i = A_1 f_{1i} + A_2 f_{2i} + \ldots + A_l f_{li} \quad (i = 1, 2, \ldots, k)$$

where $A_1, A_2, \ldots, A_k$ are arbitrary polynomials. The rows of this array are not independent.

The general case of r equations can be reduced to that of solving a single equation. Consider first the single equation

$$F_1 X_1 + F_2 X_2 + \ldots + F_k X_k = 0.$$

The conditions imposed by this equation on X_1 are merely that it must be a member of the module $(F_2, F_3, \ldots, F_k)/(F_1)$. Let $(f_{11}, f_{21}, \ldots, f_{l1})$ be a basis of this module. Then the general solution for X_1 is

$$X_1 = A_1 f_{11} + A_2 f_{21} + \ldots + A_l f_{l1}.$$

To each separate solution $X_1 = f_{j1}$ there corresponds a solution $f_{j2}, f_{j3}, \ldots, f_{jk}$ for $X_2, X_3, \ldots, X_k$, giving a row $f_{j1}, f_{j2}, \ldots, f_{jk}$ of the array of solutions. The remaining solutions are those for which $X_1 = 0$, when the equation reduces to

$$X_2 F_2 + \ldots + X_k F_k = 0.$$

To each solution for $X_2 = f_{j'2}$ $(j' = l' + 1, l' + 2, \ldots, l'')$ there corresponds a row $0, f_{j'2}, f_{j'3}, \ldots, f_{j'k}$ of the array of solutions in which the first element is zero. Similarly there are rows in which the first two elements are zero, and so on. The method may give more rows altogether than are necessary. Any row of the array which can be modified by the other rows so as to become a row of zeros should be omitted.

In the case of r equations we eliminate $X_1, X_2, \ldots, X_{r-1}$, obtaining $D_{1,2,\ldots,r} X_r + D_{1,2,\ldots,r-1,r+1} X_{r+1} + \ldots + D_{1,2,\ldots,r-1,k} X_k = 0$, and find the complete solution of this equation by the method just described. To each solution there is a unique set of values for $X_1, X_2, \ldots, X_{r-1}$ which are in general polynomials. In an exceptional case the unknowns $X_1, X_2, \ldots, X_k$ may be subjected to a linear substitution beforehand.

The principal case. The principal case is that in which the module

$$\begin{pmatrix} u_1 & u_2 & \ldots & u_k \\ v_1 & v_2 & \ldots & v_k \\ & & \ldots \ldots \ldots & \end{pmatrix}$$

is of rank $k - r + 1$. In this case it is seen from the equation in $X_r, X_{r+1}, \ldots, X_k$ above that X_k is a member of the module

$$\begin{pmatrix} u_1 & u_2 & \ldots & u_{k-1} \\ v_1 & v_2 & \ldots & v_{k-1} \\ & & \ldots \ldots \ldots \ldots & \end{pmatrix}$$

by § 53, and similarly for each unknown. The complete array of solutions is therefore obtained by putting any $k-r-1$ of the unknowns equal to zero and solving for the ratios of the remaining $r+1$ unknowns.

The $\dfrac{\lfloor k}{\lfloor r+1 \; \lfloor k-r-1}$ solutions found in this way are of the type

$$X_{p_1} = D_{p_2,\dots,p_{r+1}}, \; X_{p_2} = -D_{p_1,p_3,\dots,p_{r+1}}, \; \dots \; X_{p_{r+1}} = (-1)^r D_{p_1,p_2,\dots,p_r},$$

$$X_{p_{r+2}} = \dots = X_{p_k} = 0,$$

where $p_1, p_2, \dots, p_k$ is any permutation of $1, 2, \dots, k$.

Noether's Theorem

55. Noether's "fundamental theorem in algebraic functions" (N) furnishes a remarkably direct method of testing whether a given polynomial is a member of a given module or not; but it only attains complete success in its application to a module of rank n. A variation of the method, depending on the same principle, can be applied successfully to any module known to be primary, when the equations to its spread in the form of § 21 have been found (M, p. 88).

Noether proved that if f, ϕ were any two given polynomials in two variables x_1, x_2, without common factor, then the independent linear equations satisfied identically by the coefficients of the power products of x_1, x_2, in $A'f + B'\phi$, where A', B' are ordinary power series with undetermined coefficients, were finite and determinate; and that any polynomial F whose coefficients satisfied all these identical equations, when the origin was taken successively at each point of (f, ϕ), was a member of (f, ϕ). Thus the conditions which F has to satisfy in order to be a member of (f, ϕ) can be collected locally, so to speak, by going to each point of (f, ϕ) to find them. On going to a point not in (f, ϕ) we get no conditions, for at such a point every polynomial is of the form $A'f + B'\phi$. That the conditions are necessary is evident; for if $F = 0 \bmod (f, \phi)$ then F is of the form $A'f + B'\phi$ wherever the origin is taken.

König (K, p. 385) proved the theorem for the case of a module $(f_1, f_2, \dots, f_n)$ of rank n in n variables; and Lasker generalized the theorem in the Lasker-Noether theorem given below.

That the theorem is true for any module of rank n (not merely for a module of the principal class of rank n, the case proved by König) follows from the Hilbert-Netto and Lasker theorems. For, by Lasker's theorem, the module is the L.C.M. of a finite number of simple modules Q_1, Q_2, ..., Q_l; and if γ is the characteristic number of $Q_i = (f_1, f_2, ..., f_h)$ and the origin is taken at the point of Q_i, we have

$$F = P_1 f_1 + P_2 f_2 + ... + P_h f_h \text{ (where } P_1, P_2, ..., P_h \text{ are power series)}$$
$$= X_1 f_1 + X_2 f_2 + ... + X_h f_h \bmod O^\gamma = 0 \bmod Q_i.$$

Thus F contains $[Q_1, Q_2, ..., Q_l]$.

56. The Lasker-Noether Theorem (L, p. 95). *If*

$$M = (F_1, F_2, ..., F_k) \quad and \quad F = P_1 F_1 + P_2 F_2 + ... + P_k F_k,$$

where P_1, P_2, ..., P_k are ordinary power series, there exists a polynomial ϕ not containing the origin such that $F\phi = 0 \bmod M$.

Let Q_1, Q_2, ..., Q_l be the relevant primary modules into which M resolves, and let Q_1, Q_2, ..., $Q_{l'}$ be those which contain the origin, and $Q_{l'+1}$, ..., Q_l those which do not. Then, assuming the theorem to be true, it follows that

$$F = 0 \bmod [Q_1, Q_2, ..., Q_{l'}],$$

since ϕ cannot contain the spread of any of the modules Q_1, Q_2, ..., $Q_{l'}$. Conversely if $F = 0 \bmod [Q_1, Q_2, ..., Q_{l'}]$ and $\phi = 0 \bmod [Q_{l'+1}, ..., Q_l]$, where ϕ does not contain the origin, then $F\phi = 0 \bmod M$. Hence the aggregate of all polynomials F which are of the form

$$P_1 F_1 + P_2 F_2 + ... + P_k F_k$$

constitutes the module $[Q_1, Q_2, ..., Q_{l'}]$.

Definition. A module which resolves into primary modules all of which contain the origin, such as the module $[Q_1, Q_2, ..., Q_{l'}]$ above, will be called a *Noetherian module*.

Thus a Noetherian module, like an H-module, ceases to be such in general when the origin is changed. Moreover an H-module is a particular kind of Noetherian module; for all the primary modules into which an H-module resolves are H-modules and contain the origin.

In order that a polynomial F may be a member of a Noetherian module $(F_1, F_2, ..., F_h)$ it is sufficient that F should be of the form $P_1 F_1 + P_2 F_2 + ... + P_h F_h$.

Proof of the theorem. It is evident that the theorem is true for a module of rank n or dimensions 0 (§ 55). We shall prove the theorem for a module of dimensions $n - r$ assuming it true for a module of dimensions $n - r - 1$. It will be sufficient to prove the theorem for a primary module Q which contains the origin; for it is clear that it will then be true in general.

Let $Q = (f_1, f_2, ..., f_h)$, and $f = P_1 f_1 + P_2 f_2 + ... + P_h f_h$,

where $P_1, P_2, ..., P_h$ are power series. Let $Q_O = (f_1', f_2', ..., f_{N'}')$ be the module whose members consist of all polynomials of the form of f, and Q_P the like module obtained by moving the origin to P (and then back to O). Choose a point P so near to O as to come within the range of convergency of all the power series $P_1, P_2, ..., P_h$ for each member f_i' of the basis of Q_O when expressed in the form of f. Then we have $f_i' = 0 \bmod Q_P$, i.e. Q_O contains Q_P. But it does not follow that Q_P contains Q_O however near P may be to O; for O might be a special point of the spread of Q. We assume for the present that O is not a special point of the spread; and we choose P to be another point of the spread so near to O that Q_P contains Q_O. Then $Q_O = Q_P$.

Let u be a fixed arbitrarily chosen linear homogeneous polynomial, and f' any member of Q_O.   Then

$$f' = 0 \bmod Q_O = 0 \bmod (Q, u)_O.$$

But (Q, u) is of $n - r - 1$ dimensions; hence, assuming the general theorem as regards (Q, u), there exists a polynomial ϕ not containing O such that

$$f'\phi = 0 \bmod (Q, u) = pu \bmod Q,$$

where p is a polynomial. Hence, since $f_1', f_2', ..., f_{N'}'$ are members of Q_O,

$$f_i'\phi_i = p_i u \bmod Q \quad (i = 1, 2, ..., h');$$

hence $p_i u = 0 \bmod Q_O = 0 \bmod Q_P$;

but u does not contain P, and $\dfrac{1}{u}$ can be expanded as a power series when P is taken as origin; hence

$$p_i = 0 \bmod Q_P = 0 \bmod Q_O$$

$$= p_{i1} f_1' + p_{i2} f_2' + ... + p_{iN'} f_{N'}'.$$

Hence

$$f_i'\phi_i = (p_{i1} f_1' + p_{i2} f_2' + ... + p_{iN'} f_{N'}') u \bmod Q \quad (i = 1, 2, ..., h').$$

Solving these h' equations for $f_1', f_2', ..., f_{N'}'$ we have

$$Df_i' = 0 \bmod Q,$$

where

$$D = \begin{vmatrix} p_{11}u - \phi_1 & p_{12}u & \dots & p_{1h'}u \\ p_{21}u & p_{22}u - \phi_2 & \dots & p_{2h'}u \\ \dots & \dots & \dots & \dots \\ p_{h'1}u & p_{h'2}u & \dots & p_{h'h'}u - \phi_{h'} \end{vmatrix} = (-1)^{h'}\phi_1\phi_2\dots\phi_{h'} \bmod u.$$

Now u contains the origin, but $\phi_1, \phi_2, \dots, \phi_{h'}$ and consequently D do not; i.e. D does not contain the spread of Q. Hence $f_i' = 0 \bmod Q$. Hence Q_O contains Q, i.e., $Q_O = Q$.

This has been proved for a non-special point O of Q. If O is a special point, choose P a non-special point of Q so near to O that Q_O contains Q_P. Then since $Q_P = Q$ we have again $Q_O = Q$.

The above proof only differs from the proof given by Lasker in the part relating to $Q_O = Q_P$. In this part Lasker's proof seems to be faulty.

IV. THE INVERSE SYSTEM AND MODULAR EQUATIONS

57. A considerable number of the properties proved in this section are to be found in (M); but the introduction of the inverse system is new.

Definitions. The array of the coefficients of a complete linearly independent set of members of a module M of degree $\leqslant t$ arranged under the power products $\omega_1, \omega_2, \ldots, \omega_\mu$ of degree $\leqslant t$ is called the *dialytic array* of the module M for degree t.

The linear homogeneous equations of which this array is the array of the coefficients are called the *dialytic equations* of M for degree t.

Thus the dialytic equations of M for degree t are represented by equating all members of M of degree $\leqslant t$ to zero and regarding the power products of $x_1, x_2, \ldots, x_n$ as symbols for the unknowns.

The array inverse (§ 54) to the dialytic array of M for degree t is called the *inverse array* of M for degree t.

The linear homogeneous equations of which this array is the array of the coefficients are called the *modular equations* of M for degree t.

The modular equations for degree t are the equations which are identically satisfied by the coefficients of each and every member of M of degree $\leqslant t$. They may not be independent for members of degree $< t$ and they do not apply to members of degree $> t$ (see § 59).

The sum of the products of the elements in any row of the inverse array for degree t with the inverse power products $\omega_1^{-1}, \omega_2^{-1}, \ldots, \omega_\mu^{-1}$ is called an *inverse function* of M for degree t.

Thus the modular equations of M for degree t are represented by equating all the inverse functions of M for degree t to zero, taking each negative power product $(x_1^{p_1} x_2^{p_2} \ldots x_n^{p_n})^{-1}$ as a symbol for " the coefficient of $x_1^{p_1} x_2^{p_2} \ldots x_n^{p_n}$ in the general member of M of degree t."

We shall also say that a polynomial $F = \Sigma a_{p_1,\ldots,p_n} x_1^{p_1} \ldots x_n^{p_n}$ and a finite or infinite negative power series $E = \Sigma c_{q_1,\ldots,q_n} (x_1^{q_1} \ldots x_n^{q_n})^{-1}$ are inverse to one another if the constant term of the product FE vanishes, i.e. if $\Sigma a_{p_1, p_2, \ldots, p_n} c_{p_1, p_2, \ldots, p_n} = 0$. Thus any member of M of degree $\leqslant t$ and any inverse function of M for degree t are inverse to one another.

Any inverse function of M for degree t can be continued so as to become an inverse function of M for any higher degree (§ 59), and when continued indefinitely becomes an inverse function of M *without limitation in respect to degree.* If all coefficients after a certain stage become zero the inverse function terminates and is a finite negative power series. In the case of an H-module the inverse functions are homogeneous (§ 59) and therefore finite.

In order that a function may be an inverse function of M it is necessary and sufficient that it should be inverse to all **members** of M; hence if M contains M' any inverse function of M' is an **inverse** function of M. The whole system of inverse functions of M can therefore be resolved into primary systems corresponding to the primary modules of M. The inverse functions of a Noetherian primary module are all finite (§ 65) but not in general homogeneous. The inverse functions of a non-Noetherian primary module are all infinite power series (§ 65).

We shall regard *inverse function* and *modular equation* as convertible terms, and use that term in each case which seems best suited to the context.

A module is completely determined by its system of modular equations no less than by its system of members. The two systems are alternative representations of the module. Also the properties of the modular equations are very remarkable, and it is necessary to consider them in order to give a complete theory of modular systems.

As there is a one-one correspondence between the members of a module M of degree $\leqslant t$ and the members of the equivalent H-module of degree t, so there is a one-one correspondence between the modular equations of M *for* degree t and the modular equations of the members of the equivalent H-module of degree t. These last are called the modular equations of the H-module *of* (absolute) degree t.

58. Theorem. *The number of independent modular equations of degree t of an H-module $(F_1, F_2, \ldots, F_r)$ of rank r is the coefficient of x^t in*

$$(1 - x^{l_1})(1 - x^{l_2}) \ldots (1 - x^{l_r})(1 - x)^{-n},$$

where $l_1, l_2, \ldots, l_r$ are the degrees of $F_1, F_2, \ldots, F_r$.

Since the whole number of linearly independent polynomials of degree t is the number of power products of degree t, or the coefficient of x^t in $(1 - x)^{-n}$, the theorem will be proved if it is shown that the number $N(r, t)$ of linearly independent members of $(F_1, F_2, ..., F_r)$ of degree t is the coefficient of x^t in

$$\{1 - (1 - x^{l_1})(1 - x^{l_2}) ... (1 - x^{l_r})\}(1 - x)^{-n}.$$

This is easily seen to be true when $r = 1$.

Since any member of $(F_1, F_2, ..., F_r)$ is a linear combination of elementary members, we have

$$N(r, t) = N(r - 1, t) + \rho,$$

where ρ is the number of polynomials $\omega_1 F_r, \omega_2 F_r, ..., \omega_\rho F_r$ of degree t of which no linear combination is a member of $(F_1, F_2, ..., F_{r-1})$, or the number of power products $\omega_1, \omega_2, ..., \omega_\rho$ of degree $t - l_r$ of which no linear combination is a member of $(F_1, F_2, ..., F_{r-1})$, § 48. Hence

$$\rho + N(r - 1, t - l_r) = \text{number of power products of degree } t - l_r$$
$$= \text{coefficient of } x^t \text{ in } x^{l_r}(1 - x)^{-n};$$

and

$$N(r, t) = N(r - 1, t) - N(r - 1, t - l_r) + \text{coefficient of } x^t \text{ in } x^{l_r}(1 - x)^{-n}.$$

Hence, assuming the theorem for $N(r - 1, t)$, it follows that $N(r, t)$ is the coefficient of x^t in

$$\{1 - (1 - x^{l_1}) ... (1 - x^{l_{r-1}})\}(1 - x)^{-n}(1 - x^{l_r}) + x^{l_r}(1 - x)^{-n},$$

or in $\quad \{1 - (1 - x^{l_1}) ... (1 - x^{l_{r-1}})(1 - x^{l_r})\}(1 - x)^{-n},$

which proves the theorem.

This result is independent of the coefficients of $F_1, F_2, ..., F_r$; hence it follows that any member of $(F_1, F_2, ..., F_r)$ is expressible in one way only in the form

$$X^{(0)} F_1 + X^{(1)} F_2 + ... + X^{(r-1)} F_r,$$

where $X^{(i)}$ (as in §§ 6, 7) is a polynomial in which $x_1, x_2, ..., x_i$ occur only to powers as high as $x_1^{l_1-1}, ..., x_i^{l_i-1}$, the variables having been subjected to a substitution beforehand.

The theorem can be applied to any module $(F_1, F_2, ..., F_r)$ of rank r if $(F_1, F_2, ..., F_r)$ is an H-basis, i.e. if the H-module determined by the terms of highest degree in $F_1, F_2, ..., F_r$ is of rank r (§ 49). In this case *the number of independent modular equations for degree t is the coefficient of x^t in* $(1 - x^{l_1}) ... (1 - x^{l_r})(1 - x)^{-n-1}$. An important particular case is the following:

The number of independent modular equations of a module $(F_1, F_2, ..., F_n)$ of rank n such that the resultant of the terms of highest degree in $F_1, F_2, ..., F_n$ does not vanish is $l_1 l_2 ... l_n - 1$ for degree $l - 1$, and $l_1 l_2 ... l_n$ for any degree $\geqslant l$, where

$$l = l_1 + l_2 + ... + l_n - n.$$

This is also true for any H-module $(F_1, F_2, ..., F_n)$ of rank n ; but the number of modular equations *for* degree t will be the sum of the numbers of modular equations *of* all degrees $\leqslant t$, so that there is one modular equation of degree l and none of any degree $> l$.

59. *Any inverse function of M for any degree can be continued so as to give an inverse function of M for any higher degree.*

By carrying the continuation on indefinitely we obtain a power

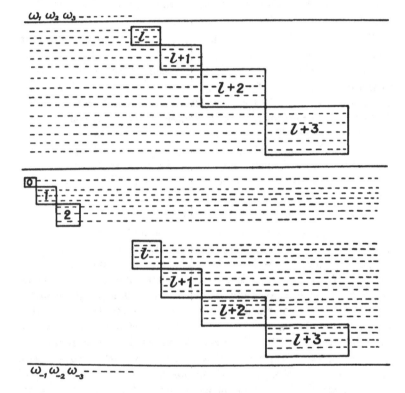

series (finite or infinite) which is an inverse function of M for all degrees without limit.

Let $(F_1, F_2, ..., F_k)$ be an H-basis of M. Then any member F of M is a linear combination of elementary members $\omega_i F_j$ *no one of which is of higher degree than F.* Let l be the lowest degree of any member of M. Write down the dialytic array of M for degree l, viz. the array of the coefficients of such members of the H-basis as are of degree l. Their terms of degree l (corresponding to the compartment l of the diagram) are linearly independent, for if not there would be a member of M of degree $< l$, which is not the case. Next write down the rows of the array representing such members of the basis as are of degree $l + 1$, and members obtained by multiplying members of degree l by $x_1, x_2, \ldots, x_n$, so as to obtain a complete set of members of degree $l + 1$ linearly independent as regards their terms of degree $l + 1$, these terms corresponding to the compartment $l + 1$ of the diagram. Proceeding in the same way we can obtain the whole dialytic array for any degree.

To obtain the inverse array for the same degree first write down square compartments $0, 1, 2, \ldots, l-1$ with arbitrary elements corresponding to degrees $0, 1, 2, \ldots, l-1$, and then a compartment l inverse to the compartment l of the dialytic array. Each row of the compartments $0, 1, 2, \ldots, l-1$ can be continued so as to be inverse to the dialytic array for degree l, since the determinants of the compartment l do not all vanish. This completes the inverse array for degree l. All its rows can be continued so as to be inverse to the dialytic array for degree $l + 1$, and a compartment $l + 1$ of new rows can be added inverse to the compartment $l + 1$ of the dialytic array. This completes the inverse array for degree $l + 1$; and we can proceed in a similar way to obtain the inverse array for any degree.

This diagram or scheme for the dialytic and inverse arrays of a given module M will be often referred to. The ease with which it can be conceived mentally is due to the fact that it is obtained by working with an H-basis of M. Each pair of corresponding compartments $l + i$ form inverse arrays, and in combination form a square array, showing that the combined complete arrays for any degree have the same number of rows as columns. In the case of a module of rank n the compartments of the dialytic array eventually become square and *the total number of rows of the inverse array is finite*. To a square compartment in either array corresponds no compartment or rows of the other array. In the case of an H-module the compartments are the only parts of the arrays whose elements do not vanish, i.e. the inverse functions are homogeneous.

Definition. The negative power series represented by the rows of the inverse array *continued indefinitely* will be called the *members of the inverse system*, and $E_1, E_2, E_3, \ldots$ will be used to denote them, just as $F_1, F_2, F_3, \ldots$ denote members of the module.

The system inverse to (1) has no member. The system inverse to $(x_1, x_2, ..., x_n)$ has only one member $E = 1$.; and the modular equation $1 = 0$ signifies that the module contains the origin.

60. Properties of the Inverse System. Before attempting to show in what ways the inverse system may be simplified we consider its general properties.

Definition. If $E = \overset{\infty}{\Sigma} c_{p_1, p_2, ..., p_n} (x_1^{p_1} x_2^{p_2} ... x_n^{p_n})^{-1}$ is a negative power series (no p_i negative), and A any polynomial, the part of the expanded product AE which consists of a negative power series will be denoted by $A.E$ and called the *A-derivate of E*. Thus

$$x_1 . (x_1 x_2)^{-1} = (x_2)^{-1}, \quad x_2^2 . (x_1 x_2)^{-1} = 0.$$

A negative power series $E = \overset{\infty}{\Sigma} c_{p_1, p_2, ..., p_n} (x_1^{p_1} x_2^{p_2} ... x_n^{p_n})^{-1}$ is or is not an inverse function of a module M according as every member $F = \Sigma a_{p_1, p_2, ..., p_n} x_1^{p_1} x_2^{p_2} ... x_n^{p_n}$ of M, or not every member of M, is inverse to it, i.e. according as every $\Sigma a_{p_1, p_2, ..., p_n} c_{p_1, p_2, ..., p_n} = 0$ or not. Suppose E an inverse function and F *any* member of M. Then $x_1^{l_1} x_2^{l_2} ... x_n^{l_n} F = \Sigma a_{p_1, p_2, ..., p_n} x_1^{p_1 + l_1} ... x_n^{p_n + l_n}$ is a member inverse to E; hence every $\Sigma a_{p_1, p_2, ..., p_n} c_{p_1 + l_1, ..., p_n + l_n} = 0$, and

$$\Sigma c_{p_1 + l_1, ..., p_n + l_n} (x_1^{p_1} x_2^{p_2} ... x_n^{p_n})^{-1},$$

or $x_1^{l_1} x_2^{l_2} ... x_n^{l_n} . E$, is a member of the inverse system. *Hence if E is a member of the inverse system of M so also is $x_1^{l_1} x_2^{l_2} ... x_n^{l_n} . E$, and if $E_1, E_2, ..., E_h$ are members so also is $A_1 . E_1 + A_2 . E_2 + ... + A_h . E_h$ a member, where $A_1, A_2, ..., A_h$ are arbitrary polynomials.*

In a slightly modified sense which will be explained later (§ 82) *the inverse system of any module M has a finite basis $[E_1, E_2, ..., E_h]$ such that any member of the inverse system is of the form*

$$X_1 . E_1 + X_2 . E_2 + ... + X_h . E_h,$$

where $X_1, X_2, ..., X_h$ are polynomials.

This theorem is evidently true in the important case in which the total number of linearly independent members of the inverse system is finite, viz. in the case of a module of rank n and in the case of a module of rank r when treated as a module in r variables only, or, in other words, in the case of a module which resolves into simple modules.

Regarding the inverse system as representing the modular equations of M we shall write $M = [E_1, E_2, ..., E_h]$ as well as $M = (F_1, F_2, ..., F_k)$. Here M is the L.C.M. of $[E_1]$, $[E_2]$, ..., $[E_h]$ and the G.C.M. of (F_1), (F_2), ..., (F_k).

Definition. A module M will be called a *principal system* if its inverse system has a basis consisting of a single member, i.e. if $M = [E]$.

A module of the principal class is a principal system (§ 72), but a principal system is not necessarily of the principal class. A principal system is however the residual of a module (F) with respect to any module of the principal class which contains, and is of the same rank as, the principal system (cf. § 62).

61. *The system inverse to $M = (F_1, F_2, ..., F_k)$ is the system whose F_i-derivates $(i = 1, 2, ..., k)$ vanish identically.*

In other words, in order that E may be a member of the inverse system of M it is necessary and sufficient that $F_i . E$ $(i = 1, 2, ..., k)$ should vanish identically. For if $E = \overset{\infty}{\Sigma} c_{q_1, q_2, ..., q_n} (x_1^{q_1} x_2^{q_2} ... x_n^{q_n})^{-1}$ is any member of the inverse system, and $F_i = \Sigma a_{p_1, p_2, ..., p_n} x_1^{p_1} x_2^{p_2} ... x_n^{p_n}$,

then $\quad F_i . E = \underset{p}{\Sigma} a_{p_1, p_2, ..., p_n} \underset{q}{\Sigma} c_{p_1 + q_1, ..., p_n + q_n} (x_1^{q_1} x_2^{q_2} ... x_n^{q_n})^{-1}$

$$= \underset{q}{\Sigma} (x_1^{q_1} x_2^{q_2} ... x_n^{q_n})^{-1} \underset{p}{\Sigma} a_{p_1, p_2, ..., p_n} c_{p_1 + q_1, ..., p_n + q_n} = 0,$$

since every $\underset{p}{\Sigma} a_{p_1, ..., p_n} c_{p_1 + q_1, ..., p_n + q_n}$ vanishes $(x_1^{q_1} ... x_n^{q_n} F_i$ being inverse to $E)$. Conversely if $F_i . E = 0$, then $\underset{p}{\Sigma} a_{p_1, ..., p_n} c_{p_1 + q_1, ..., p_n + q_n} = 0$, i.e. $x_1^{q_1} ... x_n^{q_n} F_i$ is inverse to E, and every member of M is inverse to E, i.e. E is a member of the inverse system.

Similarly if $M = [E_1, E_2, ..., E_h]$ the necessary and sufficient condition that F may be a member of M is that $F . E_j$ $(j = 1, 2, ..., h)$ vanishes identically.

62. *The modular equations of $M/(F_1, F_2, ..., F_k)$ are the F_i-derivates of the modular equations of M, i.e.*

$$[E_1, E_2, ..., E_h]/(F_1, F_2, ..., F_k) = [..., F_i . E_j, ...].$$

For the necessary and sufficient condition that F may be a member of the residual module is

$$FF_i = 0 \bmod M \quad (i = 1, 2, ..., k)$$

or $\qquad FF_i \cdot E_j = 0 \quad (i = 1, 2, ..., k; \; j = 1, 2, ..., h)$

or $\qquad\qquad F \cdot (F_i \cdot E_j) = 0.$

Hence $[..., F_i \cdot E_j, ...]$ is the residual module (§ 61).

63. *A system of negative power series with a finite basis* $[E_1, E_2, ..., E_h]$ *of such a nature that all derivates of* $E_1, E_2, ..., E_h$ *belong to the system is an inverse system of a module if* E_i $(i = 1, 2, ..., h)$ *has an* F_i-*derivate which vanishes identically.*

For there are polynomials F such that the F-derivate of each of $E_1, E_2, ..., E_h$ vanishes identically, the product $F_1 F_2 ... F_h$ being one such polynomial. Also the whole aggregate of such polynomials F constitutes a module M; for if F belongs to the aggregate so does AF. Consider the dialytic and inverse arrays of M obtained as in § 59. Since every member of M is inverse to every member of $[E_1, E_2, ..., E_h]$ all members of the latter are represented in the inverse array. If any other power series are represented, viz. if there is a row of the inverse array which does not represent a member of $[E_1, E_2, ..., E_h]$, let it begin in the compartment $l + i$. Then if we omit this row we can add a row to the dialytic array representing a polynomial of degree $l + i$ inverse to all members of $[E_1, E_2, ..., E_h]$ but not a member of M. This is contrary to the fact that M is the whole aggregate of such polynomials. Hence the system inverse to M is $[E_1, E_2, ..., E_h]$.

Thus in order that $E = \overset{\infty}{\Sigma} c_{p_1, p_2, ..., p_n} (x_1^{p_1} x_2^{p_2} ... x_n^{p_n})^{-1}$ may represent a modular equation of a module it is necessary and sufficient that $c_{p_1, p_2, ..., p_n}$ should be a recurrent function of $p_1, p_2, ..., p_n$, that is, a function satisfying some recurrent relation

$$\underset{p}{\Sigma} a_{p_1, p_2, ..., p_n} c_{p_1 + l_1, ..., p_n + l_n} = 0$$

for all positive integral values of $l_1, l_2, ..., l_n$, where the $a_{p_1, p_2, ..., p_n}$ are a set of fixed quantities finite in number. It may be that $c_{p_1, p_2, ..., p_n}$ satisfies several such recurrent relations not deducible from one another; but it is sufficient if it satisfies one.

64. *Transformation of the inverse system corresponding to a linear transformation of the modular system.*

If the variables in the modular system M are subjected to a linear non-homogeneous substitution with non-vanishing determinant by which M is transformed to M' it is required to find how the inverse system $[E_1, E_2, ..., E_h]$ is to be transformed so as to be inverse to M'.

In other words, if the negative power series E is inverse to the polynomial F it is required to find a power series E' inverse to the transformed polynomial F'. It will be shown that an E' exists which can be derived from E in a way depending only on the substitution and not on the polynomial F.

Let $\quad F = \Sigma a_{p_1, \dots, p_n} x_1^{p_1} \dots x_n^{p_n}, \quad F' = \Sigma a'_{q_1, \dots, q_n} x_1'^{q_1} \dots x_n'^{q_n},$

and let the coefficients $c_{p_1, p_2, \dots, p_n}$ of E be represented symbolically by $c_1^{p_1} c_2^{p_2} \dots c_n^{p_n}$. Then we have $E = \Sigma c_1^{p_1} \dots c_n^{p_n} (x_1^{p_1} \dots x_n^{p_n})^{-1}$; and

$$\Sigma a_{p_1, p_2, \dots, p_n} c_1^{p_1} c_2^{p_2} \dots c_n^{p_n} = 0,$$

since E, F are inverse to one another. Let the inverse substitution be $\quad x_i' = a'_{i1} x_1 + \dots + a'_{in} x_n + a_i' \quad (i = 1, 2, \dots, n).$

Then $\Sigma a'_{q_1, \dots, q_n} (a'_{11} x_1 + \dots)^{q_1} \dots (a'_{n1} x_1 + \dots)^{q_n} = \Sigma a_{p_1, \dots, p_n} x_1^{p_1} \dots x_n^{p_n},$

and we have

$\Sigma c_1^{p_1} \dots c_n^{p_n} \times$

$\quad$ {coeff. of $x_1^{p_1} \dots x_n^{p_n}$ in $\Sigma a'_{q_1 \dots q_n} (a'_{11} x_1 + \dots)^{q_1} \dots (a'_{n1} x_1 + \dots)^{q_n}$} $= 0,$

i.e. $\quad\quad\quad \Sigma a'_{q_1 \dots q_n} (a'_{11} c_1 + \dots)^{q_1} \dots (a'_{n1} c_1 + \dots)^{q_n} = 0,$

i.e. the power series $E' = \Sigma (a'_{n} c_1 + \dots)^{q_1} \dots (a'_{n1} c_1 + \dots)^{q_n} (x_1^{q_1} \dots x_n^{q_n})^{-1}$ is inverse to the polynomial $F' = \Sigma a'_{q_1, \dots, q_n} x_1^{q_1} \dots x_n^{q_n}$.

Hence the coefficient of $(x_1^{q_1} x_2^{q_2} \dots x_n^{q_n})^{-1}$ in the transformed power series E' is

$$c'_{q_1, q_2, \dots, q_n} = (a'_{11} c_1 + \dots)^{q_1} (a'_{21} c_1 + \dots)^{q_2} \dots (a'_{n1} c_1 + \dots)^{q_n},$$

where, after expanding the right-hand side, $c_1^{p_1} c_2^{p_2} \dots c_n^{p_n}$ is to be put equal to $c_{p_1, p_2, \dots, p_n}$, the coefficient of $(x_1^{p_1} x_2^{p_2} \dots x_n^{p_n})^{-1}$ in E. For such a transformation of F and E, when not inverse to one another, $\Sigma a_{p_1, p_2, \dots, p_n} c_{p_1, p_2, \dots, p_n}$ is an absolute invariant.

The most important transformation is that corresponding to a change of origin only. In this case, if

$$F = \Sigma a_{p_1 \dots p_n} x_1^{p_1} \dots x_n^{p_n} \quad \text{and} \quad E = \Sigma c_1^{p_1} \dots c_n^{p_n} (x_1^{p_1} \dots x_n^{p_n})^{-1},$$

and the new origin is the point $(-a_1, -a_2, \dots, -a_n)$,

then $\quad\quad F' = \Sigma a_{p_1 \dots p_n} (x_1 - a_1)^{p_1} \dots (x_n - a_n)^{p_n}$

and $\quad\quad E' = \Sigma (c_1 + a_1)^{p_1} \dots (c_n + a_n)^{p_n} (x_1^{p_1} \dots x_n^{p_n})^{-1}.$

It is to be noticed that if E is a finite power series it nevertheless transforms into an infinite power series E'. In particular if $E = 1$

then $E' = \Sigma a_1^{p_1} \ldots a_n^{p_n} (x_1^{p_1} x_2^{p_2} \ldots x_n^{p_n})^{-1}$, the inverse function of $(x_1 - a_1, \ldots, x_n - a_n)$.

For homogeneous substitutions another way of considering corresponding transformations of F and E can be given, which however *excludes a change of origin*. Represent E by

$$\overset{\infty}{\Sigma} c_{p_1, p_2, \ldots, p_n} \frac{u_1^{p_1} u_2^{p_2} \ldots u_n^{p_n}}{p_1! \, p_2! \ldots p_n!}$$

instead of $\Sigma c_{p_1, p_2, \ldots, p_n} (x_1^{p_1} \ldots x_n^{p_n})^{-1}$, and let the new E be defined as inverse (or conjugate) to $F = \Sigma a_{p_1, p_2, \ldots, p_n} x_1^{p_1} \ldots x_n^{p_n}$ when the same relation $\Sigma a_{p_1, \ldots, p_n} c_{p_1, \ldots, p_n} = 0$ holds as before. Then for contragredient substitutions of $x_1, x_2, \ldots, x_n$ and $u_1, u_2, \ldots, u_n$ the polynomial F and power series E will always remain inverse (or conjugate) to one another if they are so originally. Also the members E of the inverse (or conjugate) system of a module M, when expressed in the new form above, are the power series with respect to which the members (of the basis) of the module M are apolar (§ 61).

65. The Noetherian Equations of a Module.
The modular equations $\Sigma c_{p_1, p_2, \ldots, p_n} (x_1^{p_1} x_2^{p_2} \ldots x_n^{p_n})^{-1} = 0$ of a module M for degree t are finite because they are only applicable to members of degree $\leqslant t$, and the coefficients $(x_1^{p_1} x_2^{p_2} \ldots x_n^{p_n})^{-1}$ in the general member of degree t vanish when $p_1 + \ldots + p_n > t$. A modular equation may however be finite in itself, i.e. every $c_{p_1, p_2, \ldots, p_n}$ for which $p_1 + p_2 + \ldots + p_n$ exceeds a certain fixed number l may vanish. If such an equation is applied to a polynomial of degree $> l$ it only affects the coefficients of terms of degree $\leqslant l$.

Definition. The *Noetherian equations* of a module are the modular equations which are finite in themselves.

There are no Noetherian equations if the module does not contain the origin. For if $E = 0$ is a Noetherian equation of absolute degree l, and ω^{-1} a power product of absolute degree l which is present in E, the derivate equation $\omega \, . \, E = 0$ is $1 = 0$, showing that the module contains the origin. *Every Noetherian equation has the equation* $1 = 0$ *as a derivate.*

On the other hand Noetherian equations always exist if the module contains the origin, for the equation $1 = 0$ exists, and so does the equation $\omega^{-1} = 0$, where ω is any power product of less degree than any term which occurs in any member of the module.

The whole system of Noetherian equations of a non-Noetherian module M forms only a part of the whole system of modular equations, and is exhibited by a scheme similar to but different from that of § 59, with which it should be compared. In this new scheme the rows of the dialytic array represent the members of the module arranged in order according to their *underdegree* (or degree of their lowest terms) instead of their degree (or degree of their highest terms). The first set of rows represents a complete set of members of underdegree l_1 which are linearly independent as regards their terms of degree l_1,

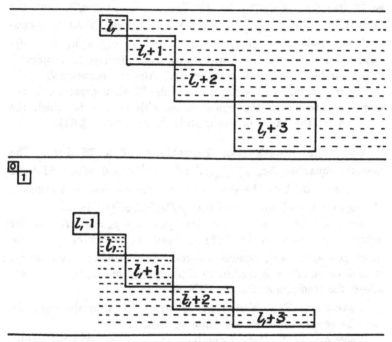

where l_1 is the lowest underdegree of any member of M. These are obtained from any basis of M, which need not be an H-basis. The next set of rows represents a complete set of members of underdegree $l_1 + 1$ which are linearly independent as regards their terms of degree $l_1 + 1$, obtained partly from the basis of M and partly from the set of members of underdegree l_1 by multiplying them by $x_1, x_2, \ldots, x_n$; and similarly for succeeding sets. The compartments $l_1, l_1 + 1, \ldots$ correspond to the terms of lowest degree in the successive sets.

To obtain the corresponding inverse (or Noetherian) array first insert square compartments $0, 1, 2, ..., l_1 - 1$ with arbitrary elements (or with elements 1 in the diagonal and the remaining elements zero) corresponding to degrees $0, 1, 2, ..., l_1 - 1$; and then a compartment l_1 inverse to the compartment l_1 of the dialytic array. This completes the array for degree l_1; all its rows are inverse to all members of M and represent Noetherian equations. Next insert a compartment $l_1 + 1$ inverse to the compartment $l_1 + 1$ of the dialytic array, and continue its rows backwards so as to be inverse to the first set of rows of the dialytic array. This completes the array for degree $l_1 + 1$; and we can proceed similarly to find in theory the whole of the Noetherian array.

The object of the diagram is merely to exhibit the whole system of Noetherian equations, which it evidently does. If F is a polynomial for which all the Noetherian equations for degree t are satisfied, then, up to and inclusive of its terms of degree t, F is a linear combination of members of the module of underdegree $\leqslant t$, i.e. F is expressible as far as degree t in the form $X_1F_1 + X_2F_2 + ... + X_kF_k$, where $X_1, X_2, ..., X_k$ are polynomials, and $F = 0 \bmod (M, O^{t+1})$. Consequently if F satisfies the whole system of Noetherian equations it is of the form $P_1F_1 + P_2F_2 + ... + P_kF_k$, where $P_1, P_2, ..., P_k$ are power series. Hence $FF_0 = 0 \bmod M$, where F_0 has a non-vanishing constant term (§ 56); and, if M is a Noetherian module, $F = 0 \bmod M$. *Hence the whole system of modular equations of a Noetherian module can be expressed as a system of Noetherian equations.*

66. Modular Equations of Simple Modules. If in the last article the rows of the compartment $l_1 + i$ of the dialytic array should be equal in number to the power products of degree $l_1 + i$ there will be no Noetherian equations of absolute degree $\geqslant l_1 + i$. In this case the Noetherian equations are finite in number and can be actually determined (at any rate in numerical examples). This can only happen when the module contains the origin as an isolated point, and the Noetherian equations are then the modular equations of the simple Noetherian module contained in the given module. The simple module itself is (M, O^{l_1+i}) and $l_1 + i$ is its characteristic number.

Thus the simple modules at isolated points of a given module M can all be found by moving the origin to each point in succession and finding its Noetherian equations and characteristic number.

Let M have a simple module at the point $(a_1, a_2, \ldots, a_n)$. Move the origin to the point and find the Noetherian equations. They will be represented by finite negative power series

$$E_1 = E_2 = \ldots = E_h = 0$$

and all derivates of the same. Also any such system represents a simple module at the origin; the fact that the coefficients of $E_1, E_2, \ldots, E_h$ are recurrent functions (§ 63) placing no restriction on them when finite in number. Let $E_i = \Sigma c_{p_1, p_2, \ldots, p_n} (x_1^{p_1} x_2^{p_2} \ldots x_n^{p_n})^{-1}$ be of absolute degree $\gamma_i - 1$. Moving the origin back to its original position, that is, to the point $(-a_1, -a_2, \ldots, -a_n)$, the equation $E_i = 0$ becomes (§ 64)

$$\Sigma (c_1 + a_1)^{p_1} (c_2 + a_2)^{p_2} \ldots (c_n + a_n)^{p_n} (x_1^{p_1} x_2^{p_2} \ldots x_n^{p_n})^{-1} = 0,$$

where, after expanding $(c_1 + a_1)^{p_1} \ldots (c_n + a_n)^{p_n}$, each $c_1^{q_1} \ldots c_n^{q_n}$ is to be put equal to the known constant $c_{q_1, q_2, \ldots, q_n}$ which it represents. Also $c_{q_1, q_2, \ldots, q_n} = 0$ if $q_1 + q_2 + \ldots + q_n \geqslant \gamma_i$. Thus

$$(c_1 + a_1)^{p_1} (c_2 + a_2)^{p_2} \ldots (c_n + a_n)^{p_n} = \left(1 + \frac{c_1}{a_1}\right)^{p_1} \ldots \left(1 + \frac{c_n}{a_n}\right)^{p_n} a_1^{p_1} a_2^{p_2} \ldots a_n^{p_n}$$

$$= k_{p_1, p_2, \ldots, p_n} a_1^{p_1} a_2^{p_2} \ldots a_n^{p_n},$$

where $k_{p_1, p_2, \ldots, p_n}$ is a whole function of $p_1, p_2, \ldots, p_n$ of degree $\gamma_i - 1$.

Hence the modular equations of any simple module at the point $(a_1, a_2, \ldots, a_n)$ are represented by power series

$$\overset{\infty}{\Sigma} k_{p_1, p_2, \ldots, p_n} a_1^{p_1} a_2^{p_2} \ldots a_n^{p_n} (x_1^{p_1} x_2^{p_2} \ldots x_n^{p_n})^{-1} = 0$$

and their derivates, where $k_{p_1, p_2, \ldots, p_n}$ is a whole function of $p_1, p_2, \ldots, p_n$. Conversely any system of equations (finite in number) of this type with all their derivates is a system of modular equations of a simple module at the point $(a_1, a_2, \ldots, a_n)$.

The following is a consequence of the above. The general solution for the recurrent function $c_{p_1, p_2, \ldots, p_n}$ (§ 63) satisfying a set of recurrent equations $\underset{p}{\Sigma} a_{p_1, p_2, \ldots, p_n} c_{p_1 + l_1, \ldots, p_n + l_n} = 0$ for all positive integral values of $l_1, l_2, \ldots, l_n$, when the corresponding polynomials $\Sigma a_{p_1, p_2, \ldots, p_n} x_1^{p_1} x_2^{p_2} \ldots x_n^{p_n}$ have only a finite number of points $(a_1, a_2, \ldots, a_n)$ in common, is $\Sigma A a_1^{p_1} a_2^{p_2} \ldots a_n^{p_n}$, where A is a whole function of $p_1, p_2, \ldots, p_n$ dependent on the point $(a_1, a_2, \ldots, a_n)$ and involving linear parameters. When the polynomials have an infinite number of points in common there can scarcely be said to be a general solution for $c_{p_1, p_2, \ldots, p_n}$.

Properties of Simple Modules

67. Theorem. *If the resultant of $(F_1, F_2, ..., F_n)$ does not vanish identically the number of Noetherian equations of any simple module of $(F_1, F_2, ..., F_n)$ is equal to the multiplicity of the corresponding solution of $F_1 = F_2 = ... = F_n = 0$ given by the resultant.*

This theorem is proved for the case $n = 2$ in (M_1, p. 388) and for the general case in (L, p. 98). Both proofs are very complicated; and a simpler proof is given here.

By the resultant of $(F_1, F_2, ..., F_n)$ we shall understand the resultant with respect to $x_1, x_2, ..., x_{n-1}$, viz. a polynomial in x_n, the variables having been subjected to a homogeneous linear substitution beforehand. Move the origin to any point of $(F_1, F_2, ..., F_n)$. Then, if x_n^C is the highest power of x_n which divides the resultant, C is the multiplicity of the solution of $F_1 = F_2 = ... = F_n = 0$ corresponding to the origin. Let Q be the whole simple module of $(F_1, F_2, ..., F_n)$ at the origin, and N the number of its modular equations. We have to prove that $C = N$.

Consider first the specially simple case in which the origin is not a singular point of the curve $(F_2, F_3, ..., F_n)$. The terms of the first degree in $F_2, F_3, ..., F_n$ are then linearly independent. For simplicity we may suppose them to be $x_2, x_3, ..., x_n$. Then F_1 can be modified by $F_2, F_3, ..., F_n$ so that its terms of lowest degree reduce to the single term x_1^p. Hence the modular equations of Q, or Noetherian equations of $(F_1, F_2, ..., F_n)$, are $x_1^{-p+1} = x_1^{-p+2} = ... = x_1^{-1} = 1 = 0$ (§ 65), so that $N = p$. Also the number of points of intersection of $F_1 = F_2 = ... = F_n = 0$ that coincide with the origin is p, so that $C = p$. Hence $C = N$.

Consider now the general case. Let $F_1', F_2', ..., F_n'$ be n polynomials whose coefficients are arbitrary except that they satisfy the N equations of Q. Then $(F_1, F_2, ..., F_n)$ and $(F_1', F_2', ..., F_n')$ have the same simple module Q at the origin, and the same N. It can be proved also that they have the same C. By the Lasker-Noether theorem (§ 56), since $(F_1, F_2, ..., F_n)$ and $(F_1', F_2', ..., F_n')$ have the same Noetherian equations, there exist polynomials $\phi_1, \phi_2, ..., \phi_n$ and $\phi_1', \phi_2', ..., \phi_n'$, none of which vanish at the origin, such that

$$\phi_i F_i = 0 \bmod (F_1', F_2', ..., F_n') \quad \text{and} \quad \phi_i' F_i' = 0 \bmod (F_1, F_2, ..., F_n).$$

Hence the module $(\phi_1 F_1, \phi_2 F_2, ..., \phi_n F_n)$ contains $(F_1', F_2', ..., F_n')$, and the resultant of the former is divisible by that of the latter (§ 11).

But the resultant of $(\phi_1 F_1, \phi_2 F_2, ..., \phi_n F_n)$ is the product of 2^n resultants of which one only, the resultant of $(F_1, F_2, ..., F_n)$, has x_n as a factor. Hence the resultant of $(F_1, F_2, ..., F_n)$ is divisible by as high a power of x_n as the resultant of $(F_1', F_2', ..., F_n')$, and *vice versa*; i.e. the two resultants are divisible by the same power of x_n.

Now the resultant of the terms of highest degree $l_1, l_2, ..., l_n$ in $F_1', F_2', ..., F_n'$ does not vanish, for the coefficients of these terms are absolutely arbitrary if $l_1, l_2, ..., l_n$ are all chosen as high as the characteristic number of Q. Hence the equations $F_1' = F_2' = ... = F_n' = 0$ have no solutions at infinity, and the number of their finite solutions is $l_1 l_2 ... l_n$, taking multiplicity into account. Also the sum of the values of N for all the points of $(F_1', F_2', ..., F_n')$ is $l_1 l_2 ... l_n$ (end of § 58), i.e. is equal to the sum of the values of C. Also each point of $(F_1', F_2', ..., F_n')$ except the origin comes under the simple case considered above; for even if the curve $(F_2', F_3', ..., F_n')$ has any singular points other than the origin, F_1' does not pass through them, since the origin is the only fixed point of F_1'. Hence the values of C and N are equal at each point of $(F_1', F_2', ..., F_n')$ other than the origin, and are therefore also equal at the origin. This proves the theorem.

68. *Definitions.* The *multiplicity* of a simple module is the number of its independent Noetherian equations.

This number has a geometrical interpretation when the theory of the resultant is applicable; but in general it has only an algebraical interpretation.

The *multiplicity of a primary module* of rank r is the multiplicity of each of the simple modules into which it resolves when regarded as a module in r variables only.

Thus there are four important numbers in connection with any primary module, viz. the rank r, the order d, the characteristic number γ, and the multiplicity μ.

A primary module of rank r will be said to be of the *principal Noetherian class* if there is a module $(F_1, F_2, ..., F_r)$ of rank r which contains it and does not contain any primary module of greater multiplicity with the same spread. On moving the origin to any general point of the spread any member of the primary module will be of the form $P_1 F_1 + P_2 F_2 + ... + P_r F_r$, where $P_1, P_2, ..., P_r$ are power series.

In other words, the primary modules into which a module of the principal class resolves are said to be of the principal Noetherian class.

Any prime module is of the principal Noetherian class; but in general a primary module is such that any module of the principal class which contains it determines a primary module of greater multiplicity. For example, O^2 is of multiplicity $n + 1$, but any module of the principal class of rank n containing O^2 contains a simple module at the origin of multiplicity 2^n at least.

If M is a module of rank n the number of its modular equations is finite and equal to the sum $\Sigma\mu$ of the multiplicities of its simple modules. In order that we may have $F = 0 \bmod M$ the coefficients of F must satisfy the $\Sigma\mu$ equations (which will not be independent unless F is of sufficiently high degree). Any set of $\Sigma\mu$ linearly independent polynomials such that no linear combination of them is a member of M is called a *complete set of remainders* for M; and has the property that any polynomial F which is not a member of M is congruent mod M to a unique linear combination of the set of remainders. The simplest way of choosing a complete set of remainders is to take the polynomial 1 of degree 0, then as many power products of degree 1 as possible, then as many power products of degree 2 as possible, and so on, till a set of $\Sigma\mu$ power products has been obtained of which no linear combination is a member of M. We shall call any such set a *simple complete set* of remainders for M.

If $M = [E_1, E_2, ..., E_h]$ is a simple Noetherian module no member E of the system $[E_1, E_2, ..., E_h]$ can have the same coefficients (assumed real) as a member F of M; for if E and F had the same coefficients the sum of their squares would be zero. Hence if the members of the system $[E_1, E_2, ..., E_h]$ have their power products changed from negative to positive they will form a complete set of remainders for M.

69. *A Noetherian principal system $[E_1]$ is uniquely expressible as a system $[E]$ such that the polynomial F with the same coefficients as E is a member of the module $[E]/O$.*

Let $E_2, E_3, ..., E_\mu$ be a complete set of linearly independent derivates of E_1 all of less absolute degree than E_1, and let $F_1, F_2, ..., F_\mu$ be the polynomials having the same coefficients as $E_1, E_2, ..., E_\mu$. Then $E_2, E_3, ..., E_\mu$ are the members of the system

$$[E_1]/O = [x_1 . E_1, x_2 . E_1, ..., x_n . E_1];$$

and $F_2, F_3, ..., F_\mu$ is a complete set of remainders for the module $[E_1]/O$. Hence there is a unique F such that

$$F = F_1 + \lambda_2 F_2 + ... + \lambda_\mu F_\mu = 0 \bmod [E_1]/O.$$

The member E of $[E_1]$ with the same coefficients as F is unique, and the system $[E]$ is the same as the system $[E_1]$. A homogeneous Noetherian equation is already in its unique form.

70. *If E is homogeneous and of absolute degree l the numbers of linearly independent derivates of E of degrees l' and $l - l'$ are equal.*

Let $E_1, E_2, \ldots, E_N$ be the members of the system $[E]$ of degree l', and $F_1, F_2, \ldots, F_L$ the members of the module $[E]$ of degree l', and $G_1, G_2, \ldots, G_N$ the polynomials which have the same coefficients as $E_1, E_2, \ldots, E_N$; so that $F_1, \ldots, F_L, G_1, \ldots, G_N$ form a complete set of linearly independent homogeneous polynomials of degree l'. Then the F_1-, F_2-, $\ldots$, F_L-derivates of E vanish identically, and the G_1-, G_2-, $\ldots$, G_N-derivates are the derivates of degree $l - l'$, and are linearly independent; otherwise some linear combination of $G_1, G_2, \ldots, G_N$ would be a member of the module $[E]$. Hence the numbers of derivates of E of degrees l' and $l - l'$ are equal.

71. *The modular equations of a simple module Q of the principal Noetherian class consist of a single equation $E = 0$ and its derivates; that is, a simple module of the principal Noetherian class is a principal system* (M, p. 109).

Take the origin at the point of Q. Then the modular equations of Q are Noetherian, and the characteristic number γ of Q is 1 more than the absolute degree of the highest modular equation. Also since Q is of the principal Noetherian class it is the whole Noetherian module contained in a certain module $M = (F_1, F_2, \ldots, F_n)$ of rank n. By choosing the degrees $l_1, l_2, \ldots, l_n$ of $F_1, F_2, \ldots, F_n$ to be $\geqslant \gamma$ we may assume $(F_1, F_2, \ldots, F_n)$ to be an H-basis of M (§ 49).

Now if F is any polynomial of degree $l_1 + l_2 + \ldots + l_n - n - 1$ such that $x_1 F, x_2 F, \ldots, x_n F$ are all members of M then F itself is a member. We prove this for 2 variables referring for the general proof to (M, p. 110). When $n = 2$, we have

$$x_1 F = A_1 F_1 + A_2 F_2, \quad x_2 F = B_1 F_1 + B_2 F_2,$$

where A_1, B_1 are of degrees $\leqslant l_2 - 2$ and A_2, B_2 of degrees $\leqslant l_1 - 2$.

Hence $\qquad x_2 (A_1 F_1 + A_2 F_2) = x_1 (B_1 F_1 + B_2 F_2),$

or $\qquad (x_2 A_1 - x_1 B_1) F_1 = (x_1 B_2 - x_2 A_2) F_2,$

or $\qquad x_2 A_1 - x_1 B_1 = 0 = x_1 B_2 - x_2 A_2,$

since $x_2 A_1 - x_1 B_1$ is of degree $< l_2$ and cannot be divisible by F_2. Hence A_1, A_2 are both divisible by x_1, and $F = 0 \bmod (F_1, F_2)$.

Suppose $Q = [E_1, E_2, ..., E_h]$, where each E_i is relevant, that is, not a member of the system $[E_1, ..., E_{i-1}, E_{i+1}, ..., E_h]$. Then the conditions that $x_1 F$, $x_2 F$, ..., $x_n F$ are to be members of M require only that the coefficients of F should satisfy all the derivates of $E_1 = E_2 = ... = E_h = 0$ (but not these equations themselves) and all the modular equations of the other simple modules of M; i.e. $l_1 l_2 ... l_n - h$ equations in all. But these conditions require $F = 0 \bmod M$, or that the coefficients of F should satisfy all the $l_1 l_2 ... l_n$ modular equations of M, which are equivalent to $l_1 l_2 ... l_n - 1$ independent equations as applied to F (§ 58). Hence the $l_1 l_2 ... l_n - h$ equations as applied to F are equivalent to no less than $l_1 l_2 ... l_n - 1$ independent equations. Hence $h = 1$, and $Q = [E_1]$.

The converse of this theorem, viz. that a simple principal system is of the principal Noetherian class, is true in the case of 2 variables (M_3), but not true in the case of more than 2 variables. Thus

$$[x_1^{-2} + x_2^{-2} + x_3^{-2}] = (x_1^2 - x_2^2, \ x_1^2 - x_3^2, \ x_2 x_3, \ x_3 x_1, \ x_1 x_2)$$

is a principal system which is not of the principal Noetherian class.

72. *A module of the principal class of rank n is a principal system.* Let $[E_1], [E_2], ..., [E_a]$ be the simple modules into which the given module resolves, and $\gamma_1, \gamma_2, ..., \gamma_a$ the characteristic numbers, and $a_1, a_2, ..., a_a$ the x_1-coordinates of the points of $[E_1], [E_2], ..., [E_a]$. The given module $[E_1, E_2, ..., E_a]$ will be proved to be identical with $[E_1 + E_2 + ... + E_a]$.

Since $x_1 - a_i$ contains the spread of $[E_i]$, $(x_1 - a_i)^{\gamma_i}$ is a member of the module $[E_i]$, § 32, and $(x_1 - a_i)^{\gamma_i} . E_i$ vanishes identically (§ 61). Hence from the equation $E_1 + E_2 + ... + E_a = 0$ we have

$$(x_1 - a_2)^{\gamma_2} (x_1 - a_3)^{\gamma_3} ... (x_1 - a_a)^{\gamma_a} . E_1 = 0.$$

The operator on the left hand is a polynomial in $x_1 - a_1$ in which the constant term does not vanish; hence if we apply the inverse operator $(x_1 - a_2)^{-\gamma_2} ... (x_1 - a_a)^{-\gamma_a}$ expanded in powers of $(x_1 - a_1)$ as far as $(x_1 - a_1)^{\gamma_1 - 1}$ we shall obtain $E_1 = 0$; since $(x_1 - a_1)^l . E_1$ vanishes identically when $l \geqslant \gamma_1$.

Hence E_1, and similarly $E_2, E_3, ..., E_a$, are all derivates of $E_1 + E_2 + ... + E_a$ and the given module $[E_1, E_2, ..., E_a] = [E_1 + E_2 + ... + E_a]$.

If M is a module of the principal class of rank r then $M^{(r)}$ and all its simple modules are principal systems. *Hence any module of the principal class, and its primary modules, are principal systems* (§ 82).

73. *If a simple module M_μ of multiplicity μ is a principal system $[E]$, and $M'_{\mu'}$ is a simple module of multiplicity μ' contained in M_μ, and $M_\mu/M'_{\mu'} = M''_{\mu''}$, then $M_\mu/M''_{\mu''} = M'_{\mu'}$, and $\mu' + \mu'' = \mu$ (M, p. 111).*

The modular equations of $M_\mu/M'_{\mu'}$ are the F-derivates of $E = 0$, where F is any member of $M'_{\mu'}$ (§ 62). Let $F_1, F_2, ..., F_{\mu'}$ be a complete set of remainders for $M'_{\mu'}$. To these can be added $F_{\mu'+1}, ..., F_\mu$ so that $F_1, F_2, ..., F_\mu$ is a complete set of remainders for M_μ. Also each of $F_{\mu'+1}, ..., F_\mu$ can be modified by a linear combination of $F_1, F_2, ..., F_{\mu'}$ so as to become a member of $M'_{\mu'}$; and we will suppose this to have been done. Then the $F_{\mu'+1}, ..., F_\mu$-derivates of $E = 0$ are modular equations of $M''_{\mu''}$, and are linearly independent, since no linear combination of $F_{\mu'+1}, ..., F_\mu$ is a member of M_μ. Also any other F-derivate of $E = 0$, where F is a member of $M'_{\mu'}$, is dependent on the $\mu - \mu'$ equations already found, since

$$F = \lambda_1 F_1 + \lambda_2 F_2 + ... + \lambda_\mu F_\mu \bmod M_\mu,$$

which requires, since $F = 0 \bmod M'_{\mu'}$,

$$\lambda_1 F_1 + \lambda_2 F_2 + ... + \lambda_{\mu'} F_{\mu'} = 0 \bmod M'_{\mu'},$$

or $$\lambda_1 = \lambda_2 = ... = \lambda_{\mu'} = 0.$$

Hence the F-derivate of $E = 0$ is the $(\lambda_{\mu'+1} F_{\mu'+1} + ... + \lambda_\mu F_\mu)$-derivate, and the number of modular equations of $M''_{\mu''}$ is $\mu - \mu'$, i.e. $\mu = \mu' + \mu''$.

Also since $M'_{\mu'} . M''_{\mu''}$ contains M_μ, $M'_{\mu'}$ contains $M_\mu/M''_{\mu''}$ which is of multiplicity $\mu - \mu'' = \mu'$. Hence $M'_{\mu'} = M_\mu/M''_{\mu''}$.

It is true in general for unmixed modules of the same rank that if M is a principal system containing M', and $M/M' = M''$, then M', M'' are mutually residual with respect to M (cf. § 24, Ex. ii).

In (M, p. 112) the opinion is expressed that if M_μ is any simple module of multiplicity μ, and $M'_{\mu'}$ any module contained in M_μ, then the multiplicity of $M_\mu/M'_{\mu'}$ cannot exceed $\mu - \mu'$. This is not correct, as the following example shows.

Example. Let
$$M_\mu = [E_1, E_2] = [(x_1 x_3)^{-1} + (x_2 x_4)^{-1}, (x_1 x_5)^{-1} + (x_2 x_6)^{-1}],$$
and $$M'_{\mu'} = (x_1, x_2, O^2) = [x_3^{-1}, x_4^{-1}, x_5^{-1}, x_6^{-1}],$$
so that $$\mu = 2 + 6 + 1 = 9, \quad \mu' = 4 + 1 = 5.$$
Then $$M_\mu/M'_{\mu'} = [E_1, E_2]/(x_1, x_2) = [x_1 . E_1, x_2 . E_1, x_1 . E_2, x_2 . E_2]$$
$$= [x_3^{-1}, x_4^{-1}, x_5^{-1}, x_6^{-1}] = M'_{\mu'}.$$
Hence (since $M_\mu/M'_{\mu'} = M'_{\mu'}$) $M'_{\mu'}$ and $M'_{\mu'}$ are mutually residual with respect to M_μ; and the multiplicity of $M_\mu/M'_{\mu'}$ is $\mu' > \mu - \mu'$.

It can be proved that if M_μ is simple and contains $M'_{\mu'}$ the multiplicity of $M_\mu/M'_{\mu'}$ cannot exceed $1 + \frac{1}{4}(\mu - \mu')^2$ or $\frac{3}{4} + \frac{1}{4}(\mu - \mu')^2$ according as $\mu - \mu'$ is even or odd.

74. *If a simple H-module M_μ of multiplicity μ is a principal system $[E]$ with characteristic number γ, and if $M'_{\mu'}$, $M''_{\mu''}$ are mutually residual with respect to M_μ, and μ_l, μ'_l, μ''_l are the numbers of linearly independent modular equations of M_μ, $M'_{\mu'}$, $M''_{\mu''}$ of degree l, then $\mu'_{l'} + \mu''_{l''} = \mu_{l'} = \mu_{l''}$, where $l' + l'' = \gamma - 1$* (M, p. 112).

Here E is homogeneous and of absolute degree $\gamma - 1$; and we have already shown that $\mu_{l'} = \mu_{l''}$ (§ 70). The $\mu''_{l''}$ modular equations of $M''_{\mu''}$ of degree l'' are F'-derivates of $E = 0$, where F' is a member of $M'_{\mu'}$ of degree l'. Hence $\mu''_{l''}$ is the number of members F' of $M'_{\mu'}$ of degree l' of which no linear combination is a member F of M_μ; for $F.E$ vanishes identically. There are $\mu_{l'}$ polynomials in all of degree l' of which no linear combination is a member of M_μ, and $\mu'_{l'}$ of these are such that no linear combination of them is a member of $M'_{\mu'}$, while the remainder $\mu_{l'} - \mu'_{l'}$ can be modified by the $\mu'_{l'}$ so as to be members of $M'_{\mu'}$. Hence

$$\mu''_{l''} = \mu_{l'} - \mu'_{l'}, \quad \text{or} \quad \mu'_{l'} + \mu''_{l''} = \mu_{l'} = \mu_{l''}.$$

Thus the values of μ''_l are known for all values of l in terms of the values of μ_l and μ'_l for all values of l.

75. *If M is any module of rank n in $x_1, x_2, \ldots, x_n$, and M_0 the equivalent H-module in $x_1, \ldots, x_n, x_0$, and μ_m the number of modular equations of $(M_0)_{x_0=0}$ of degree m, then the number of modular equations of M for degree m is*

$$H_m = 1 + \mu_1 + \mu_2 + \ldots + \mu_m.$$

This is immediately seen by considering the scheme of § 59 carried as far as degree m. The number of rows in the compartments $0, 1, 2, \ldots, l-1$ of the inverse array is the number of power products of degree $\leqslant l - 1$, and each such power product inverted represents a modular equation of $(M_0)_{x_0=0}$. This number is therefore $1 + \mu_1 + \mu_2 + \ldots + \mu_{l-1}$. The numbers of rows in the succeeding compartments are $\mu_l, \mu_{l+1}, \ldots, \mu_m$; and H_m is the total number of rows, viz. $1 + \mu_1 + \mu_2 + \ldots + \mu_m$.

Also the total number of modular equations of M, or the sum of the multiplicities of its simple modules, is equal to the multiplicity of $(M_0)_{x_0=0}$.

76. If M' is any module of rank n and μ' the sum of the multiplicities of its simple modules, we can choose n members $F_1, F_2, ..., F_n$ of M' such that the resultant of their terms of highest degree does not vanish. If then the sum of the multiplicities of the simple modules of $M = (F_1, F_2, ..., F_n)$ is μ the sum of the multiplicities of the simple modules of M/M' is $\mu - \mu'$ (§§ 71, 73), and if $M/M' = M''$ then $M/M'' = M'$. The important point is that M' is unrestricted except that it is composed of simple modules. The simple modules of M are principal systems, but not those of M'. These remarks are intended to point out the generality of the following theorem.

If $(F_1, F_2, ..., F_n)$ is an H-basis of a module M of rank n, and M' any module contained in M, and M'' the residual module M/M', then M', M'' are mutually residual with respect to M, and

$$H'_{l'} - H''_{l''} = H'_{l'+l''} - H_{l''} = H_{l'} - H''_{l'+l''},$$

where $l' + l'' + n + 1$ is the sum of the degrees of $F_1, F_2, ..., F_n$, and H_l, H'_l, H''_l are the numbers of modular equations of M, M', M'' for degree l.

This gives the values of H''_l for all values of l in terms of the values of H'_l for all values of l; for H_l is known by § 58.

The theorem is a generalization of the Brill-Noether reciprocity theorem (BN, p. 280, § 5, "Der Riemann-Roch'sche Satz"). It expresses the reciprocal relations between the numbers of the conditions which must be satisfied by members of M' and M'' in order that the product $M'M''$ may contain M.

A somewhat more general theorem is the following:

If $(F_1, F_2, ..., F_k)$ is an H-basis of a module M of rank n such that the H-module determined by the terms of highest degree in $F_1, F_2, ..., F_k$ is a principal system with characteristic number γ, and if M' is any module contained in M, and M'' the residual module M/M', then M', M'' are mutually residual with respect to M, and

$$H'_{l'} - H''_{l''} = H'_{l'+l''} - H_{l''} = H_{l'} - H''_{l'+l''}, \text{ where } l' + l'' = \gamma - 2.$$

We shall prove this more general theorem which includes the other. We must prove first that the simple modules of M are all principal systems[*]. Let M_0, M_0', M_0'' be the H-modules in $x_1, x_2, ..., x_n, x_0$ equivalent to M, M', M''. Then $(M_0)_{x_0=0}$ is a principal system; and

---

[*] The converse that if M is a module of rank n whose simple modules are all principal systems $(M_0)_{x_0=0}$ is a principal system is not true. For example, if M is the module in 2 variables determined by 3 points in a plane, then $(M_0)_{x_0=0}$ has the modular equations $x_1^{-1} = x_2^{-1} = 1 = 0$, and is not a principal system.

the multiplicities μ, μ', μ'' of $(M_0)_{x_0=0}$, $(M_0')_{x_0=0}$, $(M_0'')_{x_0=0}$ are the sums of the multiplicities of the simple modules of M, M', M'' (§ 75). Let Q' be the module determined by the a points forming the spread of M, and Q'' the residual module M/Q'. Also let Q_0', Q_0'' be the H-modules equivalent to Q', Q''. Then, since $Q'Q''$ contains M, $Q_0'Q_0''$ contains M_0, and $(Q_0'Q_0'')_{x_0=0}$ contains $(M_0)_{x_0=0}$, which is a principal system. Hence also $(Q_0'')_{x_0=0}$ contains $(M_0)_{x_0=0}/(Q_0')_{x_0=0}$ whose multiplicity is $\mu - a$; i.e. the sum of the multiplicities of the simple modules of $Q'' \geqslant \mu - a$. This is only possible when the simple modules of M are all principal systems; for if $[E_1, \ldots, E_h]$ is the simple module of M at the point P (say), the corresponding simple modules of Q', Q'' are P and $[E_1, E_2, \ldots, E_h]/P$, and the multiplicity of the latter is h less than that of $[E_1, E_2, \ldots, E_h]$; so that $\mu - \Sigma h \geqslant \mu - a$, $\Sigma h \leqslant a = a$, and $h = 1$. It follows that M' and M'' are mutually residual with respect to M.

It also follows that $\mu = \mu' + \mu''$, and that $(M_0')_{x_0=0}$ and $(M_0'')_{x_0=0}$ are mutually residual with respect to $(M_0)_{x_0=0}$. Hence $\mu'_{l'+1} + \mu''_{l''} = \mu_{l'+1} = \mu_{l''}$ (§ 74). Also $H'_{l'} = 1 + \mu_1' + \mu_2' + \ldots + \mu'_{l'}$ (§ 75). Hence

$$(H'_{l'+l''} - H'_{l'}) + H''_{l''}$$

$$= (\mu'_{l'+1} + \mu'_{l'+2} + \ldots + \mu'_{l'+l''}) + (1 + \mu_1'' + \mu_2'' + \ldots + \mu''_{l''})$$

$$= 1 + (\mu'_{l'+l''} + \mu_1'') + (\mu'_{l'+l''-1} + \mu_2'') + \ldots + (\mu'_{l'+1} + \mu''_{l''})$$

$$= 1 + \mu_1 + \mu_2 + \ldots + \mu_{l''} = H_{l''};$$

i.e. $$H'_{l'} - H''_{l''} = H'_{l'+l''} - H_{l''} = H_{l'} - H''_{l'+l''}.$$

Modular Equations of Unmixed Modules

77. We have hitherto specially considered modules of rank n, that is, modules which resolve into simple modules. The H-module of rank n is of a special type, since it is itself a simple module, and its equations are homogeneous. The general case of a module of rank n is therefore that of a module which is not an H-module. When however we consider a module of rank $< n$ it is of some advantage to replace it by its equivalent H-module, which is of the same rank but of greater dimensions by 1. We shall not avoid by this means the consideration of modules which are not H-modules, but the results obtained will be expressed more conveniently. We shall therefore assume that the given module M whose modular equations and properties are to be discussed is an H-module in n variables x_1, x_2, ..., x_n.

By treating any H-module M of rank r (whether mixed or unmixed) as a module $M^{(r)}$ in r variables $x_1, x_2, ..., x_r$ it will resolve into simple modules and have only a finite number of modular equations, viz. a number μ equal to the sum of the multiplicities of its simple modules. The unknowns in the modular equations will be represented by negative power products of $x_1, x_2, ..., x_r$ while the coefficients will be whole functions of the parameters $x_{r+1}, ..., x_n$. The module determined by these modular equations will be unmixed, viz. the L.C.M. of all the primary modules of M of rank r (§ 43); and will be the module M itself if M is unmixed. We proceed to discuss these equations and shall call them the r-*dimensional modular equations* of M (or the modular equations of $M^{(r)}$) since they are obtained by regarding the module M as a module $M^{(r)}$ in space of r dimensions. $M^{(r)}$ is not an H-module.

The dialytic array of $M^{(r)}$. We choose any basis $(F_1, F_2, ..., F_k)$ of M as the basis of $M^{(r)}$. This is not in general an H-basis of $M^{(r)}$. The module $M_{x_{r+1}=...=x_n=0}$ determined by the highest terms of the members of the basis of $M^{(r)}$ is of rank r (assuming that $x_1, x_2, ..., x_n$ have been subjected to a linear homogeneous substitution beforehand) and is therefore a simple H-module whose characteristic number will be denoted by γ.

Construct a dialytic array for $M^{(r)}$ whose elements are whole functions of $x_{r+1}, ..., x_n$ in which each row represents an elementary member $\omega_i F_j$ of $M^{(r)}$, where ω_i is a power product of $x_1, x_2, ..., x_r$ (cf. § 59). The first set of rows will represent the members of the basis which are of lowest degree l, the next set a complete set of elementary members of degree $l + 1$ which are linearly independent of one another and of the complete rows in the first set, the next set a complete set of elementary members of degree $l + 2$ linearly independent of one another and of the complete rows in the first two sets, and so on.

In comparing this with the scheme of § 59 there is the obvious difference that the elements of the array are whole functions of $x_{r+1}, ..., x_n$ instead of pure constants; and there is the more important difference that *the compartments $l, l + 1, ... do not necessarily consist of independent rows*, because the array is not constructed from an H-basis of $M^{(r)}$. It is only the complete rows of the array that are independent. *The elements in the compartments are all pure constants independent of $x_{r+1}, ..., x_n$*. The diagram of § 59 serves perfectly well to illustrate the dialytic array although its properties are now different.

In each compartment we choose a set of independent rows such that all the remaining rows of the compartment are dependent on them, and we name them *regular* rows and *extra* rows respectively, and apply the same terms to the complete rows of which they form part. In the compartment γ the regular rows will form a square array, and the same will be true of the compartments $\gamma + 1$, $\gamma + 2$, ... Eventually a compartment $\delta \geqslant \gamma$ will be reached such that the number of rows in the whole array for degree δ is exactly μ less than the whole number of columns, where μ is the number of modular equations of $M^{(r)}$ as mentioned above. After this all succeeding compartments $\delta + 1$, $\delta + 2$, ... will consist of square arrays only without any extra rows.

We can now modify any extra row of the array by regular rows so as to make all its elements which project beyond the columns of degree $\gamma - 1$ vanish, and this leaves its elements in the columns up to degree $\gamma - 1$ whole functions of $x_{r+1}, \ldots, x_n$ of the same degrees as they were before. If this is done with all the extra rows projecting beyond the columns of degree $\gamma - 1$ the array may be said to be brought to its *regular form* in which the whole number of rows of the array for degree $\gamma - 1$ is μ less than the whole number of columns, and all the compartments γ, $\gamma + 1$, ... are made square. The extra rows, modified so as to end at the columns of degree $\gamma - 1$, represent members of $M^{(r)}$ of degree $\gamma - 1$ which are not elementary members $\omega_i F_j$.

We may further modify the *regular form* of the complete array for degree $\gamma - 1$ so as to reduce the number of rows in each compartment $\gamma - 1$, $\gamma - 2$, ... successively to independent rows. The elements of some of the rows of the array for degree $\gamma - 1$ may thus become fractional in $x_{r+1}, \ldots, x_n$, and the whole number of compartments will in general be increased, so that the last (or first) compartment will be numbered $l' < l$. Supposing this to be done we can choose a *simple* complete set of remainders for $M^{(r)}$ consisting of all power products of $x_1, x_2, \ldots, x_r$ of degree $< l'$ and as many power products of each degree $l'' \geqslant l'$ as the number of columns of the compartment l'' exceeds the number of rows of the same. We denote these power products in ascending degree by $\omega_1, \omega_2, \ldots, \omega_\mu$ (so that $\omega_1 = 1$) and all remaining power products to infinity in ascending degree by $\omega_{\mu+1}, \omega_{\mu+2}, \ldots$. The two series $\omega_1, \omega_2, \ldots, \omega_\mu$ and $\omega_{\mu+1}, \omega_{\mu+2}, \ldots$ overlap in respect to the degrees of their terms.

The basis of M used for constructing the dialytic array of $M^{(r)}$ must be one in which each member is of the same degree in $x_1, x_2, \ldots, x_r$ as in $x_1, x_2, \ldots, x_n$. We shall say that M is a *perfect* module if the array of $M^{(r)}$ as originally constructed has no extra rows, i.e. if the basis $(F_1, F_2, ..., F_k)$ is an H-basis of $M^{(r)}$.

78. Solution of the dialytic equations of $M^{(r)}$**.** We return to what has been called above the regular form of the dialytic array of $M^{(r)}$. Each row represents a member of $M^{(r)}$ and supplies a congruence equation mod $M^{(r)}$. Solving these equations, regarding $\omega_{\mu+1}, \omega_{\mu+2}, ...$ as the unknowns, we have

$$D\omega_p + D_{p1}\omega_1 + D_{p2}\omega_2 + ... + D_{p\mu}\omega_\mu = 0 \bmod M^{(r)} \quad (p = \mu+1, \mu+2, ...).$$

There are two slightly different cases according as the degree of $\omega_p < \gamma$ or $\geqslant \gamma$. If ω_p is of degree $< \gamma$ we use the regular form of the array for degree $\gamma - 1$ for solving for ω_p. D is then the determinant of this array formed from the columns corresponding to $\omega_{\mu+1}, \omega_{\mu+2}, ...$, and D_{pi} the determinant formed from the columns corresponding to $\omega_{\mu+1}, ..., \omega_{p-1}, \omega_i, \omega_{p+1}, ...$. If ω_p is of degree $\geqslant \gamma$ we must use the array up to the degree of ω_p in order to solve for ω_p. D is the same as in the former case except for a factor independent of $x_{r+1}, ..., x_n$ (since the compartments $\gamma, \gamma+1, ...$ are square and all their elements are pure constants) by which the equation can be divided. Also D_{pi} is a sum of products of determinants of the regular form of the array for degree $\gamma - 1$ with determinants from the remaining rows of the larger array, so that the H.C.F. of the determinants of the array for degree $\gamma - 1$ can be divided out, and we obtain in both cases

(A) $R\omega_p + R_{p1}\omega_1 + ... + R_{p\mu}\omega_\mu = 0 \bmod M^{(r)} \quad (p = \mu+1, \mu+2, ...).$

This equation is homogeneous in $x_1, x_2, ..., x_n$, and each R_{pi} is homogeneous in $x_{r+1}, ..., x_n$. Also, owing to the fact that the remainders $\omega_1, \omega_2, ..., \omega_\mu$ are a *simple* set, each ω_p is congruent mod $M^{(r)}$ to a linear combination of those power products $\omega_1, \omega_2, ..., \omega_\mu$ which are of equal or less degree than ω_p. *Hence R_{pi} vanishes if the degree of ω_i exceeds the degree of ω_p. Also $R = 1$ if M is perfect* (cf. § 81).

79. The modular equations of $M^{(r)}$**.** If the coefficient of $\omega_p = x_1^{p_1} x_2^{p_2} ... x_r^{p_r}$ in the general member of $M^{(r)}$ of any degree is represented by $\omega_{-p} = (x_1^{p_1} x_2^{p_2} ... x_r^{p_r})^{-1}$ we have

$$\omega_{-1}\omega_1 + \omega_{-2}\omega_2 + ... + \omega_{-p}\omega_p + ... = 0 \bmod M^{(r)},$$

and, by (A),

$$R(\omega_{-1}\omega_1 + ... + \omega_{-\mu}\omega_\mu) = \sum_{p=\mu+1}^{\infty} \omega_{-p}(R_{p1}\omega_1 + R_{p2}\omega_2 + ... + R_{p\mu}\omega_\mu) \bmod M^{(r)}.$$

Here coefficients of $\omega_1, \omega_2, ..., \omega_\mu$ on both sides are equal, i.e.

(B) $$R\omega_{-i} = \sum_{p=\mu+1}^{\infty} R_{pi}\omega_{-p} \quad (i = 1, 2, ..., \mu).$$

This is the complete system of modular equations of $M^{(r)}$, or r-dimensional modular equations of M, and the system includes all its own derivates. R and all the R_{pi} are definite whole functions of $x_{r+1}, ..., x_n$. If any other complete system were given and solved for $\omega_{-1}, \omega_{-2}, ..., \omega_{-\mu}$ in terms of $\omega_{-\mu-1}, \omega_{-\mu-2}, ...$ the result would be the unique system (B).

Since in (A) $R\omega_p$ and $R_{pi}\omega_i$ are of the same degree in $x_1, x_2, ..., x_n$, so in (B), $R\omega_{-i}$ and $R_{pi}\omega_{-p}$ are of the same degree, i.e. all terms in one equation (B) are of the same degree in $x_1, x_2, ..., x_n$. Also since (§ 78) R_{pi} vanishes if the degree of ω_i exceeds the degree of ω_p *there is no ω_{-p} on the right-hand side of* (B) *of less absolute degree than ω_{-i}*; but every ω_{-p} of the same degree as ω_{-i} and not among $\omega_{-1}, \omega_{-2}, ..., \omega_{-\mu}$ will appear on the right-hand side of (B).

(B) is the complete system of r-dimensional equations of the L.C.M. of all the primary modules of M of rank r; and will decompose into separate distinct systems corresponding to the separate primary modules of rank r if M has more than one irreducible spread of rank r.

The n-dimensional equations. We can obtain the whole system of n-dimensional equations of M corresponding to the system (B) as follows: ω_{-p} or $(x_1^{p_1} x_2^{p_2} ... x_r^{p_r})^{-1}$ represents the whole coefficient of $x_1^{p_1} x_2^{p_2} ... x_r^{p_r}$ in the general member of $M^{(r)}$, i.e. it stands for

$$\Sigma (x_1^{p_1} ... x_n^{p_n})^{-1} x_{r+1}^{p_{r+1}} ... x_n^{p_n},$$

the summation extending to all values of $p_{r+1}, ..., p_n$ only. If this be substituted for each $(x_1^{p_1} ... x_r^{p_r})^{-1}$ in each of the equations (B) the whole coefficients of the power products of $x_{r+1}, ..., x_n$ will represent the n-dimensional equations. *This will be the whole system of n-dimensional equations of M if M is unmixed, as we shall assume hereafter is the case.*

The whole system of modular equations of a mixed module may be regarded as consisting of the separate systems corresponding to the primary modules into which it resolves.

80. *The system of homogeneous equations*

(C) $$R\omega_{-i} = \Sigma R_{pi}\omega_{-p} \quad (i = 1, 2, ..., \mu)$$

obtained from the system (B) *by retaining only those terms on the right hand in which R_{pi} and ω_{-p} are of the same degrees as R and ω_{-i}*

respectively is the complete system of equations of the simple H-module determined by the highest terms in x_1, x_2, ..., x_r of the members of an H-basis of $M^{(r)}$.

This can be seen by considering the diagram of § 59 assuming that it had been constructed from an H-basis of $M^{(r)}$. The compartments l, $l+1$, $l+2$, ... in the two arrays in § 59 are the dialytic and inverse arrays of the simple H-module determined by the highest terms of the members of the H-basis; and the modular equations of this simple H-module are represented by the compartments 0, 1, ..., l, $l+1$, ... of the inverse array. The system (C) is that which is represented by the compartments of the inverse array.

81. *If $R=1$ the module M (assumed unmixed) is perfect.* Since M is unmixed every whole member of $M^{(r)}$ is a member of M (§ 43). Also, since $R=1$, there is an inverse array of $M^{(r)}$ each of whose compartments consists of independent rows in which all the elements are pure constants. Hence there is a corresponding dialytic array having the same property. From this it follows that M is perfect (§ 77).

82. The r-dimensional and n-dimensional equations of M.

If the system (B) is a principal system, i.e. if all its equations are derivates of a single one of them, each simple module of $M^{(r)}$ is a principal system; for if F is a polynomial containing all the simple modules of $M^{(r)}$ except one, then $M^{(r)}/(F)$ is the last one, and is a principal system (§ 62). The converse is also true (see § 72). Also the unmixed module M in n variables is a principal system, as we proceed to prove.

Let the r-dimensional equation of which all the equations of the system (B) are derivates be

$$\sum_{}^{\infty} R_{p_1, p_2, ..., p_r} (x_1^{p_1} x_2^{p_2} ... x_r^{p_r})^{-1} = 0,$$

where $R_{p_1, p_2, ..., p_r}$ is a homogeneous polynomial in x_{r+1}, ..., x_n of degree $p_1 + p_2 + ... + p_r + \delta$. The integer δ may be negative, but the more unfavourable case for the proof is that in which it is positive. Let $c_{p_1, p_2, ..., p_n}$ be the coefficient of $x_{r+1}^{p_{r+1}} ... x_n^{p_n}$ in $R_{p_1, p_2, ..., p_r}$, so that $p_{r+1} + ... + p_n = p_1 + ... + p_r + \delta$. To convert the equation into an n-dimensional equation we put

$$(x_1^{p_1} x_2^{p_2} ... x_r^{p_r})^{-1} = \sum_q^{\infty} x_{r+1}^{q_{r+1}} ... x_n^{q_n} (x_1^{p_1} ... x_r^{p_r} x_{r+1}^{q_{r+1}} ... x_n^{q_n})^{-1}$$

as in § 79, and we have

$$\underset{p}{\Sigma} c_{p_1, \dots, p_n} x_{r+1}^{p_{r+1}} \dots x_n^{p_n} \underset{q}{\Sigma} x_{r+1}^{q_{r+1}} \dots x_n^{q_n} (x_1^{p_1} \dots x_r^{p_r} x_{r+1}^{q_{r+1}} \dots x_n^{q_n})^{-1} = 0, \dots (1)$$

or, equating the whole coefficient of $x_{r+1}^{l_{r+1}} \dots x_n^{l_n}$ to zero,

$$\underset{p}{\Sigma} c_{p_1, p_2, \dots, p_n} (x_1^{p_1} \dots x_r^{p_r} x_{r+1}^{l_{r+1}-p_{r+1}} \dots x_n^{l_n-p_n})^{-1} = 0, \dots\dots (2)$$

which is homogeneous and of absolute degree $l_{r+1} + \dots + l_n - \delta$. Similarly the general n-dimensional equation obtained from the coefficient of $x_{r+1}^{m_{r+1}} \dots x_n^{m_n}$ in the $x_1^{t_1} \dots x_r^{t_r}$-derivate of (1) is

$$\underset{p}{\Sigma} c_{p_1, p_2, \dots, p_n} (x_1^{p_1-t_1} \dots x_r^{p_r-t_r} x_{r+1}^{m_{r+1}-p_{r+1}} \dots x_n^{m_n-p_n})^{-1} = 0, \dots\dots (3)$$

where $t_1, \dots, t_r, m_{r+1}, \dots, m_n$ are any n fixed positive integers (including zeros) such that $t_1 + \dots + t_r \leqslant$ a fixed limit τ (since there are only a finite number of linearly independent derivates of the original r-dimensional equation) and $(x_1^{p_1-t_1} \dots x_r^{p_r-t_r} x_{r+1}^{m_{r+1}-p_{r+1}} \dots x_n^{m_n-p_n})^{-1}$ is zero if any one of the indices $p_1 - t_1, \dots, p_r - t_r, m_{r+1} - p_{r+1}, \dots, m_n - p_n$ is negative.

Consider all the n-dimensional modular equations of degree l, that is, all the equations of the system (3) of absolute degree l. The absolute degree of (3) is

$$m_{r+1} + \dots + m_n - \delta - t_1 - \dots - t_r = l.$$

Hence each of $m_{r+1}, \dots, m_n$ is equal to or less than $l + \delta + \tau$; and every equation (3) of absolute degree l is a derivate of the single equation (2) if $l_{r+1}, \dots, l_n$ are all chosen as high as $l + \delta + \tau$. Hence there is a single equation of which all the modular equations of M of degree l are derivates, and any equation (2) in which $l_{r+1}, \dots, l_n$ are not numerically specified will serve for the single equation.

It follows that *the inverse system of any module M has a finite basis* $[E_1, E_2, \dots, E_h]$; for M resolves into a finite number of primary modules of the same or of different ranks, and each primary module of rank r has a finite number of r-dimensional equations, and a smaller number of r-dimensional equations of which all the others are derivates, and an equal or still smaller number of n-dimensional equations of which all the others are derivates.

83.   If (B) is a principal system it does not follow that (C) is a principal system (footnote § 76). *If however* (C) *is a principal system*

(B) *is a principal system.* For the basis equation of the system (C) must be the homogeneous equation

$$R\omega_{-\mu} = \Sigma R_{p\mu}\omega_{-p},$$

and all the other equations of (C) must be of less absolute degree. Now the system (B) is unique and any equation obtained from it

$$R_1\omega_{-1} + R_2\omega_{-2} + \ldots + R_p\omega_{-p} + \ldots = 0$$

must be the result of multiplying the equations of (B) by $R_\mu, R_{\mu-1}, \ldots, R_1$ and adding and dividing out R. Hence the equation

$$R\omega_{-i} = \overset{\infty}{\Sigma} R_{pi}\omega_{-p}$$

is exactly the same derivate of $R\omega_{-\mu} = \overset{\infty}{\Sigma} R_{p\mu}\omega_{-p}$ as the corresponding homogeneous equation $R'\omega_{-i} = \Sigma R_{pi}\omega_{-p}$ is of $R\omega_{-\mu} = \Sigma R_{p\mu}\omega_{-p}$.

If (C) is a principal system the formulae of § 76 apply to any two modules M', M'' mutually residual with respect to M when regarded as modules in r variables. If (B) is a principal system, but not (C), the formula $\mu = \mu' + \mu''$ applies, where μ, μ', μ'' are the numbers of equations in the systems (B), (B'), (B'') for M, M', M''. This follows from § 73 by summing for all the simple modules of $M^{(r)}$.

84. Modular equations of an *H*-module of the principal class.

In the case of an *H*-module $(F_1, F_2, \ldots, F_r)$ of rank r (C) is a principal system (§ 71); and $R = 1$, since $(F_1, F_2, \ldots, F_r)$ is an *H*-basis of $M^{(r)}$ (§ 49). Also, if F_1, F_2, $\ldots$, F_r are of degrees $l_1, l_2, \ldots, l_r$, a complete set of remainders for $M^{(r)}$ consists of the $l_1 l_2 \ldots l_r$ factors of $x_1^{l_1-1} x_2^{l_2-1} \ldots x_r^{l_r-1}$, since this is a complete set of remainders for $(x_1^{l_1}, x_2^{l_2}, \ldots, x_r^{l_r})$, cf. § 58. Hence the system (B) for M consists of the single equation

$$(x_1^{l_1-1} \ldots x_r^{l_r-1})^{-1} = \overset{\infty}{\Sigma} R_{p_1, p_2, \ldots, p_r} (x_1^{p_1} x_2^{p_2} \ldots x_r^{p_r})^{-1} \quad\ldots\ldots(4)$$

and its derivates, where $p_1 + \ldots + p_r \geqq l_1 + \ldots + l_r - r$, and consequently $p_i \geqq l_i$ for one value at least of i. The corresponding n-dimensional equation is (§ 82)

$$\underset{q}{\Sigma} x_{r+1}^{q_{r+1}} \ldots x_n^{q_n} (x_1^{l_1-1} \ldots x_r^{l_r-1} x_{r+1}^{q_{r+1}} \ldots x_n^{q_n})^{-1}$$

$$= \underset{p}{\Sigma} c_{p_1, \ldots, p_n} x_{r+1}^{p_{r+1}} \ldots x_n^{p_n} \underset{q}{\Sigma} x_{r+1}^{q_{r+1}} \ldots x_n^{q_n} (x_1^{p_1} \ldots x_r^{p_r} x_{r+1}^{q_{r+1}} \ldots x_n^{q_n})^{-1},$$

or, by equating coefficients of $x_{r+1}^{l_{r+1}-1} \ldots x_n^{l_n-1}$ on both sides,

$$(x_1^{l_1-1} x_2^{l_2-1} \ldots x_n^{l_n-1})^{-1} = \underset{p}{\Sigma} c_{p_1, p_2, \ldots, p_n} (x_1^{p_1} \ldots x_r^{p_r} x_{r+1}^{l_{r+1}-1-p_{r+1}} \ldots x_n^{l_n-1-p_n})^{-1}.$$

When F_1, F_2, ..., F_r are general with letters for coefficients, the $c_{p_1, p_2, ..., p_n}$ are rational functions of the coefficients and on multiplying up by their common denominator K we can write the equation

$$K (x_1^{l_1-1} x_2^{l_2-1} ... x_n^{l_n-1})^{-1} = \underset{p}{\Sigma} K_{p_1, p_2, ..., p_n} (x_1^{p_1} x_2^{p_2} ... x_n^{p_n})^{-1}, \quad ...(5)$$

where $p_1 + p_2 + ... + p_n = l_1 + l_2 + ... + l_n - n$, and at least one $p_i \geqslant l_i$ $(i = 1, 2, ..., r)$ and every $p_j < l_j$ $(j = r + 1, ..., n)$. This is the n-dimensional modular equation of $(F_1, F_2, ..., F_r)$ of which all others are derivates, $l_{r+1}, ..., l_n$ being unspecified numerically. More explicitly it is the unique modular equation of the simple module $(F_1, F_2, ..., F_r, x_{r+1}^{l_{r+1}}, ..., x_n^{l_n})$; for it is a relation satisfied by the coefficients of the general member of $(F_1, F_2, ..., F_r)$ of degree $l_1 + l_2 + ... + l_n - n$ in which $p_j < l_j$ $(j = r + 1, ..., n)$, i.e. it is the unique relation (§ 58) satisfied by the coefficients of the general member of

$$(F_1, ..., F_r, x_{r+1}^{l_{r+1}}, ..., x_n^{l_n})$$

of degree $l_1 + ... + l_n - n$. The coefficients $K_{p_1, p_2, ..., p_n}$ are whole functions of the coefficients of F_1, F_2, ..., F_r of a similar kind to the resultant of $(F_1, ..., F_r, x_{r+1}^{l_{r+1}}, ..., x_n^{l_n})$ and of degree 1 less than this resultant in the coefficients of each of F_1, F_2, ..., F_r, viz. of degree $L_i - 1$ in the coefficients of F_i where $L_i l_i = l_1 l_2 ... l_n = L$. The vanishing of $K_{p_1, p_2, ..., p_n}$ is the condition that

$$x_1^{p_1} x_2^{p_2} ... x_n^{p_n} = 0 \bmod (F_1, F_2, ..., F_r, x_{r+1}^{l_{r+1}}, ..., x_n^{l_n})$$

(§ 61, since the $x_1^{p_1} x_2^{p_2} ... x_n^{p_n}$-derivate of (5) then vanishes), whereas the non-vanishing of the resultant is the condition that every power product of degree $l_1 + ... + l_n - n + 1$ is a member of the module. It is probable that some of the quantities $K_{p_1, p_2, ..., p_n}$ factorise but that they have not all a common factor. The resultant of

$$(F_1, F_2, ..., F_r, x_{r+1}^{l_{r+1}}, ..., x_n^{l_n})$$

is $R_{r+1}^{l_{r+1} ... l_n}$ (§ 8).

85. Whole basis of the system inverse to $M^{(r)}$. The simplest *whole* basis $[E_1, E_2, ..., E_h]$ of the r-dimensional system inverse to an unmixed H-module M of rank r, or the simplest expression for the system of equations (B), satisfies the following conditions: (i) each E_i $(i = 1, 2, ..., h)$ is a *whole* member of the inverse system, i.e. its coefficients are whole functions of the parameters $x_{r+1}, ..., x_n$;

(ii) all the members E_1, E_2, ..., E_h are relevant ; (iii) any whole member of the system $[E_1, E_2, ..., E_h]$ is of the form

$$X_1 . E_1 + X_2 . E_2 + ... + X_h . E_h,$$

where X_1, X_2, ..., X_h are whole functions of x_{r+1}, ..., x_n as well as of x_1, x_2, ... x_r; (iv) E_1, E_2, ..., E_h have as high absolute under-degrees in x_1, x_2, ..., x_r as possible. A whole basis, as distinguished from a simplest whole basis, is defined by (i) and (iii).

A basis $(F_1, F_2, ..., F_k)$ of M furnishes a whole basis of $M^{(r)}$, and any whole basis of $M^{(r)}$ satisfying the condition corresponding to (iii) above is a basis of M. A simplest whole basis* of $M^{(r)}$ is one in which the degrees of $F_1, F_2, ..., F_k$ in $x_1, x_2, ..., x_r$ are as low as possible.

If $(F_1, F_2, ..., F_r)$ is any module of rank r containing M such that $(F_1, F_2, ..., F_r)_{x_{r+1}=...=x_n=0}$ is of rank r, and $M = (F_1, F_2, ..., F_{k'})$, and the degrees of F_{r+1}, ..., $F_{k'}$ in $x_1, x_2, ..., x_r$ are as low as possible, the basis $(F_1, F_2, ..., F_{k'})$ will be called *a whole basis of $M^{(r)}$ in reference to* $(F_1, F_2, ..., F_r)$. All of F_{r+1}, ..., $F_{k'}$ are to be relevant, but some or all of $F_1, F_2, ..., F_r$ may be irrelevant for a basis of M.

86. Properties of H-modules mutually residual with respect to an H-module of the principal class.

Let $F_1, F_2, ..., F_r$, of degrees $l_1, l_2, ..., l_r$, be any r members of the unmixed H-module M of rank r such that

$$(F_1, F_2, ..., F_r)_{x_{r+1}=...=x_n=0}$$

is of rank r; and let M' be the residual module $(F_1, F_2, ..., F_r)/M$. Also let $(F_1, ..., F_r, F'_{r+1}, ..., F'_{r+h})$ be a whole basis of $M'^{(r)}$ in reference to $(F_1, F_2, ..., F_r) = [E]$. Since F'_{r+i} is of as low degree in $x_1, x_2, ..., x_r$ as possible the terms of F'_{r+i} of highest degree in $x_1, x_2, ..., x_r$ do not form a member of the module

$$(F_1, F_2, ..., F_r)_{x_{r+1}=...=x_n=0},$$

and are therefore of degree $l'_{r+i} \leqslant l_1 + l_2 + ... + l_r - r$ in $x_1, x_2, ..., x_r$. Also, since E begins with terms which represent the modular equation

---

* A simplest whole basis of $M^{(r)}$ is a whole basis which approaches most nearly to an H-basis; but is not necessarily an H-basis. For example,

$$(x_1^4, x_1^3 x_2, x, x_2^3, x_2^4, x_1^3 x_3^2 + x_1^2 x_2^2 x_4, x_2^3 x_4^2 + x_1^2 x_2^2 x_3)$$

is the basis of a module M of rank 2, and a simplest whole basis of $M^{(2)}$, but not an H-basis of $M^{(2)}$; since $x_1^3 x_3^3 - x_2^3 x_4^3$ is needed for an H-basis of $M^{(2)}$, but is irrelevant for a basis of M or whole basis of $M^{(2)}$.

of $(F_1, F_2, ..., F_r)_{x_{r+1}=...=x_n=0}$ of degree $l_1 + l_2 + ... + l_r - r$, F'_{r+i}. E will begin with terms of absolute degree $l_1 + ... + l_r - r - l'_{r+i}$ in $x_1, x_2, ..., x_r$ which do not vanish identically.

Now M, M' are mutually residual with respect to $(F_1, ..., F_r)$ or $[E]$. Hence

$$M = [E]/M' = [E]/(F_1, ..., F_r, F'_{r+1}, ..., F'_{r+h})$$
$$= [F'_{r+1} . E, F'_{r+2} . E, ..., F'_{r+h} . E].$$

This basis of the r-dimensional system inverse to M is a simplest whole basis $[E_1, E_2, ..., E_h]$ as defined in § 85. All its members are relevant, for if (say)

$$F'_{r+h} . E = (X_1 F'_{r+1} + ... + X_{h-1} F'_{r+h-1}) . E,$$

then

$$F'_{r+h} - X_1 F'_{r+1} - ... - X_{h-1} F'_{r+h-1} = 0 \bmod (F_1, F_2, ..., F_r),$$

which is not the case. Also any r-dimensional modular equation of M is a derivate of $E = 0$, and if a whole equation, is $F'. E = 0$, where F' is a whole function of $x_1, x_2, ..., x_n$, since $[E]$ is a whole basis; and if F is any member of M, $FF'. E$ vanishes identically, i.e.

$$FF' = 0 \bmod (F_1, F_2, ..., F_r)$$

and

$$F' = 0 \bmod M' = X_1 F'_{r+1} + ... + X_h F'_{r+h} \bmod (F_1, F_2, ..., F_r),$$

and

$$F''. E = X_1 . E_1 + X_2 . E_2 + ... + X_h . E_h.$$

Finally the absolute underdegrees of $E_1, E_2, ..., E_h$ are as high as possible since the degrees of $F'_{r+1}, ..., F'_{r+h}$ in $x_1, x_2, ..., x_r$ are as low as possible. The coefficients of the terms in E_i and F'_{r+i} which involve the parameters $x_{r+1}, ..., x_n$ to the least degree involve them to the same degree, so that $E_1, E_2, ..., E_h$ and $F'_{r+1}, F'_{r+2}, ..., F'_{r+h}$ are of the same degree of complexity in this respect.

It follows from the above that if M' is the residual of a given unmixed H-module M of rank r with respect to any H-module $(F_1, F_2, ..., F_r)$ of rank r containing M, and if

$$M' = (F_1, F_2, ..., F_r, F'_{r+1}, ..., F'_{r+h}),$$

where $F'_{r+1}, ..., F'_{r+h}$ are all relevant, then h is a fixed number independent of the choice of $F_1, F_2, ..., F_r$, viz. the number of members in a simplest whole basis $[E_1, E_2, ..., E_h]$ of the system inverse to $M^{(r)}$. Also if the degrees of $F'_{r+1}, ..., F'_{r+h}$ in respect to $x_1, x_2, ..., x_r$ are made as low as possible the degree of F'_{r+i} in respect to $x_1, x_2, ..., x_r$

is $l - a_i$ and in respect to x_1, x_2, ..., x_n is $l - a_i + \beta_i$, where l is the sum of the degrees of F_1, F_2, ..., F_r diminished by r, a_i is the absolute degree of the terms with which E_i begins, and β_i is the degree of the coefficients of these terms in x_{r+1}, ..., x_n.

87. The Theorem of Residuation. As in the last article let M be any unmixed H-module of rank r, and $(F_1, F_2, ..., F_r)$ any module of rank r containing M, and M' the residual module $(F_1, F_2, ..., F_r)/M$, so that M, M' are mutually residual with respect to $(F_1, F_2, ..., F_r)$. In geometrical terminology M, M' are residuals on $(F_2, F_3, ..., F_r)$ determined by the section F_1. Replace F_1 by another member F_1' of M, which we will suppose to be of the same degree as F_1, giving another section of $(F_2, F_3, ..., F_r)$ through M, and let $M_1' = (F_1', F_2, ..., F_r)/M$ be the residual section or module.

Also let F' be a section through M', F' being of the same degree as F_1, and $M_1 = (F', F_2, ..., F_r)/M'$ the residual section or module. Then M', M_1' are coresidual on $(F_2, F_3, ..., F_r)$ having a common residual M; and M_1 is any other residual of M'. *The theorem of residuation says that every*

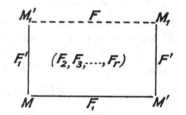

residual M_1 of M' on $(F_2, F_3, ..., F_r)$ is also a residual of M_1', i.e. to every section F' through M' there corresponds a section F through M_1' having the same residual section on $(F_2, F_3, ..., F_r)$. This theorem is a generalization of Sylvester's theory of residuation (Salmon's *Higher Plane Curves*, Chap. v) and the *Restsatz* of Brill and Noether* (BN, p. 271). Besides this relation of M' to M_1' there are properties connecting them both with M which are proved in the last article, viz. the number of members over and above F_1, F_2, ..., F_r (or F_1', F_2, ..., F_r) required for a basis of M' (or M_1') is equal to the number of members required for a whole basis of the system inverse to $M^{(r)}$; and the number of members required for a whole basis of the system inverse to $M'^{(r)}$ (or $M_1'^{(r)}$) is equal to the number of members over and above F_1, F_2, ..., F_r (or F_1', F_2, ..., F_r) required for a basis of M.

---

* It would be more correct to say that the *Restsatz* can be deduced from the theorem proved here; but not such extensions of it as have been made to surfaces etc., because these bring in mixed modules. The module M may be composed of *any* primary modules of rank r; and corresponding to each one which is not of the principal Noetherian class M' must contain a residual primary module with the same spread.

The polynomials F_1, F_1', F' and the modules M, M', M_1, M_1' having been defined as above it is required to prove that there exists a polynomial F such that M_1, M_1' are mutually residual with respect to $(F, F_2, ..., F_r)$. Let μ, μ', μ_1, μ_1' be the numbers of r-dimensional modular equations of M, M', M_1, M_1'; then

$$\mu + \mu' = \mu + \mu_1' = \mu' + \mu_1 = l_1 l_2 ... l_r$$

and therefore each equals $\mu_1 + \mu_1'$. Let ϕ, ϕ', ϕ_1, ϕ_1' be general members of M, M', M_1, M_1' with coefficients involving linear parameters. Then

$$F'F_1' = 0 \bmod MM' = 0 \bmod (F_1, F_2, ..., F_r) = FF_1 \bmod (F_2, ..., F_r),$$
$$......(1)$$

where F is a polynomial of the same degree as F_1, F_1', F'. Also

$$F_1 \phi_1' = 0 \bmod MM_1' = X_1'F_1' \bmod (F_2, ..., F_r), \quad ......(2)$$

and $\qquad \phi \phi_1' = 0 \bmod MM_1' = XF_1' \bmod (F_2, ..., F_r)$;

hence by cross multiplying and dividing out $\phi_1'F_1'$,

$$X_1'\phi = XF_1 \bmod (F_2, ..., F_r) = 0 \bmod (F_1, F_2, ..., F_r),$$
$$\therefore X_1' = 0 \bmod (F_1, F_2, ..., F_r)/M = 0 \bmod M'.$$

Similarly $\quad F_1\phi_1 = 0 \bmod M'M_1 = X'F' \bmod (F_2, ..., F_r), \quad ......(3)$

where $X' = 0 \bmod M$;

$$\therefore X'X_1' = 0 \bmod MM' = X_1 F_1 \bmod (F_2, ..., F_r). \quad ......(4)$$

Multiplying (1), (2), (3), (4), and dividing out $F'F_1'F_1^2X'X_1'$,

$$\phi_1\phi_1' = X_1 F \bmod (F_2, ..., F_r) = 0 \bmod (F, F_2, ..., F_r).$$

Hence M_1M_1' contains $(F, F_2, ..., F_r)$; and since M_1, M_1' have only μ_1, μ_1' r-dimensional modular equations, while $(F, F_2, ..., F_r)$ has $\mu_1 + \mu_1'$ and is a principal system, it follows that $M_1^{(r)}$, $M_1'^{(r)}$, and consequently M_1, M_1', are mutually residual with respect to $(F, F_2, ..., F_r)$.

The theorem has been proved on the supposition that the modules are H-modules and the degrees of F_1, F_1', F' are equal; but it is true without any of these restrictions. In the case of modules which are not H-modules the region at infinity must be regarded as non-existent and the usual conception of residual and coresidual must be slightly extended. Thus if through a point P on a plane cubic curve two lines are drawn parallel to two asymptotes cutting the curve again in Q, R, then P is residual to Q and R, and Q, R are coresidual. If through Q a line is drawn cutting the curve again in two points these two are residual to R, i.e. a curve (viz. a conic) can be drawn through them and R which does not meet the curve again except at infinity.

As an illustration of the general theorem we may suppose M to be any unmixed module of rank 2 in space of three dimensions. Then F_1, F_2 are any two surfaces containing M whose whole intersection consists of a finite number of irreducible spreads (excluding infinity); and to each spread or curve corresponds a primary principal system of (F_1, F_2). M contains a certain part of some of these principal systems and no part of others; M' has to contain the whole of the latter and the residual part of each of the former. These conditions determine M', and similarly for M_1 and M_1'.

88. Perfect Modules. *Definition.* If a module M of rank r in n variables and the corresponding module $M^{(r)}$ in r variables have a common H-basis of which each member is of the *same degree in the r variables as in the n variables* then M is called a *perfect module.*

Any module of rank n is perfect, by definition.

An unmixed H-module of rank $n-1$ is perfect; for its basis is an H-basis of $M^{(n-1)}$.

An H-module of the principal class is perfect (§ 49).

A module of the principal class which is not an H-module is not necessarily perfect. For example, the module $(x_1^2,\ x_2 + x_1 x_3)$ whose H-basis is $(x_1^2,\ x_1 x_2,\ x_2^2,\ x_2 + x_1 x_3)$, § 38, is not perfect since $x_2 + x_1 x_3$ is of less degree in x_1, x_2 than in x_1, x_2, x_3.

A prime module is not necessarily perfect. For example, the prime module of rank 2 whose spread is given by $u = \lambda u_1 = \lambda^3 u_3 = \lambda^4 u_4$, where u, u_1, u_3, u_4 are linear, has an H-basis

$$(uu_4 - u_1 u_3,\ u_1^3 - u^2 u_3,\ u_1^2 u_4 - uu_3^2,\ u_1 u_4^2 - u_3^3) = (f, f_1, f_2, f_3)$$

and no other member than $uu_4 - u_1 u_3$ of degree 2. But it has a second member $\lambda_1 f_1 + \lambda_2 f_2 + \lambda_3 f_3 + (\lambda_4 x_1 + \lambda_5 x_2)f$ which can be made of degree 2 in x_1, x_2; hence it is not perfect.

89. *An H-module M of rank r is perfect or not according as the multiplicity of the simple module $M_{x_{r+1}=\dots=x_n=0}$ is equal to or greater than the number of modular equations of $M^{(r)}$ or of $M^{(r)}{}_{x_{r+2}=\dots=x_n=0}$.* The difference between the two numbers when M is unmixed is the total number of *extra* rows of the dialytic array of $M^{(r)}$ when carried as far as degree δ (§ 77), and when M is mixed is still greater. The property affords the simplest test for deciding whether a given module is perfect or not; for the two numbers can generally be found. For example, the prime module M in § 88 is of rank 2 and order 4, while the multiplicity of $M_{x_3=\dots=x_n=0}$ is 5, so that M is not perfect. The property may also be stated in the form that *an H-module M of rank r is perfect or not according as $M^{(r)}{}_{x_{r+2}=\ldots=x_n=0}$ is perfect (i.e. unmixed) or not.*

90. *A perfect module is unmixed.* If M is perfect the module $M^{(r)}$ has an H-basis of which each member has its highest terms independent of the parameters $x_{r+1}, \ldots, x_n$. Hence the dialytic array of $M^{(r)}$ constructed from an H-basis has pure constants for the elements in all its compartments; and a non-vanishing determinant D can be selected from the array for any degree t which is a pure constant. Let ϕF be a member of M, where ϕ is a whole function of the parameters only. Then F is a member of $M^{(r)}$ and if we insert a row in the array representing F it will be dependent on the rest, i.e.

$$F = \lambda_1 F_1 + \lambda_2 F_2 + \ldots + \lambda_\rho F_\rho,$$

where $F_1, F_2, \ldots, F_\rho$ are the members of $M^{(r)}$ represented by the rows of the array, and $\lambda_1, \lambda_2, \ldots, \lambda_\rho$ are rational functions of $x_{r+1}, \ldots, x_n$. Equating coefficients on the two sides of power products of $x_1, x_2, \ldots, x_r$ corresponding to the columns of the determinant D mentioned above, and solving for $\lambda_1, \lambda_2, \ldots, \lambda_\rho$, we see that $\lambda_i D$ and consequently λ_i is a whole function of $x_{r+1}, \ldots, x_n$. Hence F is a member of M, since $F_1, F_2, \ldots, F_\rho$ are all members of M; and $\phi F = 0 \bmod M$ requires $F = 0 \bmod M$. Hence M is unmixed.

If M is a perfect module of rank r and M' a module in $x_{r+1}, \ldots, x_n$ (independent of $x_1, x_2, \ldots, x_r$) the L.C.M. of M, M' is the same as their product MM'. For if the F above is a member of the L.C.M. of M, M' the elements in the row representing F are all members of M', and the λ_i are linear functions of them and therefore also members of M'. Hence

$$F = \Sigma \lambda_i F_i = 0 \bmod MM', \quad \text{i.e.} \quad [M, M'] = MM',$$

since $\lambda_i = 0 \bmod M'$ and $F_i = 0 \bmod M$.

91. *The number H_l of modular equations of degree l of a perfect H-module M of rank r is the coefficient of x^l in*

$$(1 + \mu_1 x + \mu_2 x^2 + \ldots + \mu_{\gamma-1} x^{\gamma-1})(1-x)^{r-n},$$

where γ is the characteristic number, and μ_m the number of modular equations of degree m, of the simple module $M_{x_{r+1}=\ldots=x_n=0}$.

For the general member of M of degree l is (§ 90)

$$\lambda_1 F_1 + \lambda_2 F_2 + \ldots + \lambda_\rho F_\rho,$$

where $\lambda_1, \lambda_2, \ldots, \lambda_\rho$ are whole functions of $x_{r+1}, \ldots, x_n$, and cannot vanish identically unless $\lambda_1, \lambda_2, \ldots, \lambda_\rho$ all vanish identically. Hence the number of linearly independent members of M of degree l is the total number of terms in λ_1, λ_2, ..., λ_ρ. Now the number of the polynomials F_1, F_2, ..., F_ρ which are of degree m is μ_m less than the number of power products of x_1, x_2, ..., x_r of degree m, and the number of terms in each corresponding λ (of degree $l-m$) is the coefficient of x^l in $x^m (1-x)^{r-n}$. Hence the number of linearly independent members of M of degree l is less than the number of power products of x_1, x_2, ..., x_n of degree l by the coefficient of x^l in

$$(1 + \mu_1 x + \mu_2 x^2 + \ldots + \mu_l x^l)(1 - x)^{r-n} ;$$

and this coefficient is the value of H_l. § 75 is a particular case.

92. *If M is a perfect H-module of rank r such that the simple module $M_{x_{r+1}=\ldots=x_n=0}$ is a principal system, and M' a perfect H-module of rank r contained in M, the module M/M' is perfect.*

The μ and μ' r-dimensional modular equations of M and M' begin with the μ and μ' modular equations of $M_{x_{r+1}=\ldots=x_n=0}$ and $M'_{x_{r+1}=\ldots=x_n=0}$. Also the $\mu - \mu'$ r-dimensional modular equations of M/M' are the F'-derivates of the modular equations of M, where F' is any member of M', and begin with the $F'_{x_{r+1}=\ldots=x_n=0}$-derivates of the modular equations of $M_{x_{r+1}=\ldots=x_n=0}$, that is, with the modular equations of $M_{x_{r+1}=\ldots=x_n=0}/M'_{x_{r+1}=\ldots=x_n=0}$. These are $\mu - \mu'$ in number, since $M_{x_{r+1}=\ldots=x_n=0}$ is a principal system containing $M'_{x_{r+1}=\ldots=x_n=0}$. Hence M/M' is perfect (§ 81).

93. We may sum up some of the relations between different kinds of modules.

A module of the principal class is unmixed and a principal system, and in the case of an H-module is perfect.

Any power of a module of the principal class is unmixed, and in the case of an H-module is perfect (§ 89, end), but is not a principal system ; e.g. $(x_1, x_2)^2$ is not a principal system.

A module of rank $k - r + 1$ whose basis is a matrix with r rows and k columns is unmixed, and in the case of an H-module is perfect (§ 89, end), but is not a principal system ; e.g. the module $\begin{pmatrix} 0 & x_1 & x_2 \\ x_1 & x_2 & 0 \end{pmatrix}$ is not a principal system.

A primary module of the principal Noetherian class is a principal system, but not perfect.

NOTE ON THE THEORY OF IDEALS

The following is a brief explanation of the theory of ideals of algebraic numbers[*] and functions and the relation in which the theory given in the preceding pages stands with respect to it.

Gauss (*Disquisitiones Arithmeticae* (1801)) was the first to consider the laws of factorisation in a domain of whole numbers other than that of rational whole numbers $0, \pm 1, \pm 2, \dots$. He proved that two given complex whole numbers $a+b\sqrt{-1}$, $c+d\sqrt{-1}$ (a, b, c, d rational integers) have always an H.C.F. and that any such number is a unique product of prime factors. Kummer (*J. reine angew. Math.* 35 (1847), 40 (1850), 53 (1857)) in extending the research to a larger class of whole numbers found that these properties were no longer absolutely true. Nevertheless he succeeded in making such numbers amenable to all the simpler laws of rational integers by introducing certain *ideal numbers* not existing in the domain considered; and thus laid the foundation of the theory of factorisation of whole algebraic numbers. Finally Dedekind (D), by using *ideals* instead of ideal numbers, extended the theory to the whole numbers of any algebraic corpus and to whole algebraic functions of one variable (DW); while Kronecker (Kr) extended the same theory of factorisation to algebraic functions in general. Kronecker went still further; he gave the first steps of a *general* theory of ideals of algebraic functions (Kr, p. 77) under the name of *modular systems*. In this general theory factorisation plays only a subsidiary part, since an ideal which is not prime is not in general a product of prime ideals.

Modules of whole rational functions (as defined pp. 1, 2 above) are ideals and modules in the sense of Dedekind; and the theory of such modules is the necessary starting point of the general theory of ideals.

An *algebraic number* is any root a of an algebraic equation

$$a_0 x^m + a_1 x^{m-1} + \dots + a_m = 0$$

---

* The following are notable general accounts of the theory of algebraic numbers:

D. Hilbert. "Bericht über die Theorie der algebraischen Zahlkörper" (*Jahresb. d. deutschen Math.-Verein.*, Berlin (1897), Bd. IV).

H. Weber. *Lehrbuch der Algebra* (Brunswick, 2nd ed. (1899), Bd. II, p. 553).

G. B. Mathews. "Number" (*Ency. Brit.*, Cambridge, 11th ed. (1911), Vol. 19, p. 847).

For other references to the arithmetic theory of algebraic numbers and functions see (D), (DW), (K), and (Kr), p. xiii.

in which the coefficients $a_0, a_1, ..., a_m$ are rational integers. We may suppose that a_0 is positive, and that $a_0, a_1, ..., a_m$ have no common factor other than 1, and that the equation is irreducible in the rational domain. There is only one set of values of $a_0, a_1, ..., a_m$ satisfying these conditions for an assigned a.

a is called a *whole* (algebraic) number if $a_0 = 1$, and is called a *fractional* number if $a_0 \neq 1$. Thus an algebraic (as well as a rational) number is integral or fractional, but cannot be both. In any case $a_0 a$ is a whole number β, and $a = \beta/a_0$, i.e. the denominator of any fractional algebraic number a can be rationalized, while the numerator remains a whole algebraic number β.

All roots of any equation $x^n + c_1 x^{n-1} + ... + c_n = 0$ (whether reducible or not) in which $c_1, c_2, ..., c_n$ are rational integers are whole algebraic numbers. For all irreducible factors of the left-hand side are of the type

$$x^m + a_1 x^{m-1} + ... + a_m.$$

We omit the proof of this as of most other properties to be stated. Hence any number is whole if it satisfies any equation of this type.

If $a, \beta, \gamma, ...$ are whole numbers $a \pm \beta$ and $a\beta$ are also whole numbers (D, p. 145); and so also is any whole rational function of $a, \beta, \gamma, ...$ with rational integral coefficients.

A whole number a is said to have another β as a factor, or to be divisible by β, if $a = \beta\gamma$, where γ is a whole number.

A whole number ϵ is called a *unit* if it is a factor of 1; or ϵ is a unit if ϵ and $1/\epsilon$ are both whole numbers. Thus if in the above equation $a_0 = \pm a_m = 1$ all its roots are units.

Two whole numbers a, β are said to be *equivalent* (as regards divisibility) if $a = \epsilon\beta$ where ϵ is a unit; for then any whole number which divides either a or β divides the other. Such equivalence of a, β is denoted by $a \sim \beta$.

A *corpus* of algebraic numbers is the aggregate of all rational functions (with rational coefficients) of any finite set of given algebraic numbers $a_1, a_2, ..., a_k$. All numbers of the corpus are rational functions of a single element

$$a = c_1 a_1 + c_2 a_2 + ... + c_k a_k,$$

where $c_1, c_2, ..., c_k$ are rational integers so chosen as not to be connected by special relations.

The corpus generated by a is denoted by $\Omega(a)$ and the aggregate of algebraic integers included in the corpus by $\omega(a)$. The *order* of the corpus and of a is the degree of the irreducible equation of which a is a root.

Thus $\Omega(1)$ is the corpus of rational numbers and $\omega(1)$ the aggregate of rational integers.

Any rational function of any finite number of elements of $\Omega(a)$ is an element of $\Omega(a)$, and any whole rational function with rational integral coefficients of any finite number of elements of $\omega(a)$ is an element of $\omega(a)$.

Any corpus $\Omega(a)$ includes $\Omega(1)$, for $a/a = 1$.

If $a_0 x^m + a_1 x^{m-1} + ... + a_m = 0$ is the irreducible equation of which the element a of the corpus $\Omega(a)$ is a root, the other roots $a', a'', ..., a^{(m-1)}$ are called the *conjugates* of a, and $\Omega(a'), \ldots, \Omega(a^{(m-1)})$ the conjugates of $\Omega(a)$. If a' is an element of $\Omega(a)$ then $\Omega(a')$ is the same as $\Omega(a)$, and if not, not. The corpus generated by $a, a', \ldots, a^{(m-1)}$ is called the Galoisian domain corresponding to $\Omega(a)$. The conjugates of any number $\beta = f(a)$ of $\Omega(a)$ are

$$\beta' = f(a'), \ldots, \beta^{(m-1)} = f(a^{(m-1)}).$$

The product $\beta \beta' \ldots \beta^{(m-1)}$ is a rational number (being a symmetric function of $a, a', \ldots, a^{(m-1)}$) and is called the *norm* of β and written norm β.

Since β and norm β are both numbers in $\Omega(a)$, norm β/β is a number in $\Omega(a)$. Moreover if β is a number in $\omega(a)$, then $\beta', \ldots, \beta^{(m-1)}$ are whole algebraic numbers, and norm β/β is a number in $\omega(a)$. If β is a unit, $\beta', \beta'', \ldots, \beta^{(m-1)}$ are all algebraic units, and norm $\beta = \pm 1$. Conversely, if β is a number in $\omega(a)$ such that norm $\beta = \pm 1$, β is a unit in $\omega(a)$.

Norm $(au + \beta v + \ldots)$ is defined as $\prod\limits_{i=0}^{m-1} (a^{(i)}u + \beta^{(i)}v + \ldots)$, $u, v, \ldots$ being indeterminates.

$\Omega(a)$ is a domain of rationality. $\omega(a)$ is called a *proper holoid* domain (König), that is, a domain in which every sum, difference and product, but not every quotient, of two elements is an element of the domain. A proper holoid domain in which every pair of elements a, β have an H.C.F. in the domain (defined as a factor δ of a and of β such that every common factor of a, β is a factor of δ) is called a *complete* holoid domain. $\omega(a)$ is not necessarily *complete*.

The simplest example of this last statement is the domain $\omega(\sqrt{-5})$ which is fully discussed by Dedekind (D, p. 73). If $x = a + b\sqrt{-5}$ (a, b rational) then $(x-a)^2 + 5b^2 = 0$, and in order that x may be whole $2a$ and $a^2 + 5b^2$ must be rational integers, i.e. a and b must be integers. Consider the two whole numbers $9, 3(1 + \sqrt{-5})$. If these have an H.C.F. in $\omega(\sqrt{-5})$ it must be 3δ, where δ is a whole number in $\omega(\sqrt{-5})$ which divides 3 and $1 + \sqrt{-5}$. But 3 and $1 + \sqrt{-5}$ are non-factorisable in $\omega(\sqrt{-5})$; hence $\delta = 1$. Hence the H.C.F. (if any) of 9, $3(1 + \sqrt{-5})$ is 3; but $2 - \sqrt{-5}$ is a factor of 9 and $3(1 + \sqrt{-5})$ and is not a factor of 3. Hence there is no H.C.F., and $\omega(\sqrt{-5})$ is not *complete*. That 3 is not factorisable in $\omega(\sqrt{-5})$ is shown by putting $3 = (a + b\sqrt{-5})(c + d\sqrt{-5})$ from which it follows that

$$9 = (a^2 + 5b^2)(c^2 + 5d^2)$$

and that one of $a^2 + 5b^2$, $c^2 + 5d^2$ is 9 and the other 1, since neither can be 3. Also if $a^2 + 5b^2 = 9$ the only solutions are $a = \pm 3$, $b = 0$ and $a = \pm 2$, $b = \pm 1$ of which the latter must be rejected since $\pm 2 \pm \sqrt{-5}$ does not divide 3. Similarly for $1 + \sqrt{-5}$. The numbers $3, 1 + \sqrt{-5}$ have however a common factor $(1 + \sqrt{-5})/\sqrt{2}$ or $\sqrt{-2 + \sqrt{-5}}$ not in $\omega(\sqrt{-5})$.

Another point requiring notice is the distinction between a non-factorisable number and a prime number. A non-factorisable number in $\omega(a)$ is one which has no other factors in $\omega(a)$ than such as are equivalent to itself or 1. A prime number is one which cannot be a factor of a product $\beta\gamma$ without being a factor of β or of γ. Thus 3 and $1+\sqrt{-5}$ are both non-factorisable in $\omega(\sqrt{-5})$, but neither of them is prime; 3 is a factor of $(1+\sqrt{-5})(1-\sqrt{-5})$ but not a factor of $1+\sqrt{-5}$ or $1-\sqrt{-5}$; and $1+\sqrt{-5}$ is a factor of 6 but not a factor of 2 or 3.

In a complete holoid domain every non-factorisable element is prime and every prime element is non-factorisable (K, p. 15).

Let π be any element of the domain which is non-factorisable, and let $\alpha\beta$ be divisible by π and β not divisible by π. Then the H.C.F. of β, $\pi \sim 1$; H.C.F. of $\alpha\beta$, $\alpha\pi \sim \alpha$; H.C.F. of $\alpha\beta$, $\pi \sim$ H.C.F. of $\alpha\beta$, $\alpha\pi$, $\pi \sim$ H.C.F. of α, π; i.e. H.C.F. of α, $\pi \sim \pi$, or α is divisible by π; hence π is prime. Again if π is prime and equal to $\pi_1\pi_2$, one of π_1, π_2 is divisible by π and the other is a unit; hence π is non-factorisable.

It is to be noticed that the proof depends only on the notions of product, quotient, and H.C.F., and is therefore applicable to any domain in which each pair of elements α, β has a product $\alpha\beta$, and an H.C.F. δ (defined as above), and may or may not have a quotient γ, defined by $\alpha=\beta\gamma$.

In a complete holoid domain any element which is not an infinite product of factors (not counting unit factors) is a unique product of prime factors if equivalent factors are regarded as the same factor.

For any element which is not prime is a product of two factors neither of which is a unit; each of these again if not prime is a product of two factors, and so on. Hence any element which is not an infinite product is a product of prime factors $p_1^{l_1} p_2^{l_2} \dots p_r^{l_r}$. This resolution into factors is unique in the sense of equivalence; for if $p_1^{l_1} \dots p_r^{l_r} \sim q_1^{m_1} \dots q_s^{m_s}$, where $q_1, q_2, \dots, q_s$ are primes and none of them units, p_1 must be a factor of q_1 or q_2 ... or q_s, and if a factor of q_1, then $p_1 \sim q_1$; from which the rest follows.

The domain of all algebraic integers is a complete holoid domain (D, p. 247) but contains no prime numbers, since any number a has an infinite number of factors, e.g. $\sqrt[n]{a}$. This property of completeness is peculiar to numbers; it does not hold for functions, not even for relatively whole algebraic functions of a single variable.

No number in $\omega(a)$ can be an infinite product, for otherwise its norm, which is a rational integer, would be an infinite product of rational integers. Hence if $\omega(a)$ is complete each number in it is a unique product of prime factors.

All the above remarks concerning algebraic numbers (with the exception noted) apply *mutatis mutandis* to algebraic functions. The only difference is that there are two kinds of whole algebraic functions, relative and absolute.

An *algebraic function* is any quantity a which satisfies an algebraic equation

$$A_0 z^m + A_1 z^{m-1} + \dots + A_m = 0$$

in which the coefficients $A_0, A_1, \dots, A_m$ are whole rational functions of n variables $x_1, x_2, \dots, x_n$.

a is called a *relatively whole* (algebraic) function if A_0 does not involve the variables. In this case the numerical coefficients of $A_1, A_2, \ldots, A_m$ may be any real or complex numbers, whether algebraic or not. Moreover z is whole relatively to $x_1, x_2, \ldots, x_r$ if A_0 involves $x_{r+1}, \ldots, x_n$ only.

a is called an *absolutely whole* (algebraic) function if $A_0 = 1$ and the numerical coefficients of $A_1, A_2, \ldots, A_m$ are rational integers.

In the case of functions $\Omega(1)$ is the corpus of rational functions, and $\omega(1)$ the aggregate of whole rational functions.

In still continuing to speak of algebraic numbers it will be understood that what is said applies equally to algebraic functions. In considering the properties of algebraic numbers we naturally regard the numbers of a corpus $\Omega(a)$ and the domain $\omega(a)$ included in it as the principal subject of investigation, since these answer the most nearly to the numbers of $\Omega(1)$ and $\omega(1)$.

$\Omega(a)$ and more especially $\omega(a)$ are further subdivided. Dedekind defines a *module* in $\Omega(a)$ to be the aggregate of all numbers (or functions)

$$a_1 a_1 + a_2 a_2 + a_3 a_3 + \ldots,$$

where $a_1, a_2, \ldots$ are fixed elements of $\Omega(a)$ and $a_1, a_2, \ldots$ any elements of $\omega(1)$, that is, rational integers in the case of number modules and whole rational functions (relative or absolute) in the case of function modules. If $a_1, a_2, \ldots$ are whole numbers, that is, elements of $\omega(a)$ instead of $\Omega(a)$, the module is a module of whole numbers. Any module of whole numbers (or functions) has a finite basis $(\mu_1, \mu_2, \ldots, \mu_k)$; and any module of fractions with a finite basis $(a_1, a_2, \ldots, a_k)$ is practically the same thing as a module of whole numbers, since $a_1, a_2, \ldots, a_k$ can be multiplied by a rational integer a so as to become whole numbers $\mu_1, \mu_2, \ldots, \mu_k$, and then any element of the module $(a_1, a_2, \ldots, a_k)$ is equal to the corresponding element of the module $(\mu_1, \mu_2, \ldots, \mu_k)$ divided by a. There are modules of fractions with infinite bases; but they seem to be unimportant, and it would be simpler to restrict the meaning of the term module to a module of whole numbers or functions. A *module* would then be defined as any aggregate of elements of $\omega(a)$ such that if a_1, a_2 are any two elements of the module, $a_1 + a_2$ and $a a_1$ are also elements of the module, where a is any element of $\omega(1)$.

An *involution* of whole functions is any aggregate of elements of $\omega(a)$ such that if a_1, a_2 are any two elements of the involution, $a_1 + a_2$ and $c a_1$ are also elements of the involution, where c is any constant. In the absolute theory the elements of $\omega(a)$ are absolutely whole functions and c a rational integer.

Dedekind's definition of an *ideal* is similar but still more fundamental. An ideal is any aggregate of elements of $\omega(a)$ such that if a_1, a_2 are any two elements of the ideal, $a_1 + a_2$ and μa_1 are also elements of the ideal, where μ is any element of $\omega(a)$. Every ideal has a finite basis $(a_1, a_2, \ldots, a_k)$ and is a finite module $(a_1, a_2, \ldots, a_l)$; but not every module of whole numbers or functions is an ideal. In the domain of whole rational functions an ideal and a module are identical.

Kummer had found that the integers of a corpus did not necessarily satisfy all the simple laws of rational integers, or in other words they need not form a complete holoid domain. It occurred to Dedekind (and apparently independently to Kronecker) to consider in this case not the individual integers of a corpus only but sets of integers. For this purpose Dedekind made use of the ideals already defined. We shall in the first place consider ideals from a rather abstract point of view. The remarks apply also to some extent to modules and involutions.

The aggregate of elements of an ideal $(a_1, a_2, ..., a_k)$ constitutes an image of the properties possessed in common by all the elements of the aggregate, and especially of properties of divisibility (if any) common to $a_1, a_2, ..., a_k$. The term *ideal* should strictly be applied to these properties common to all the elements, whatever they may be ; but it is more convenient and concise to define the ideal as the aggregate of elements itself. This point of view, viz. that the ideal is a set of properties rather than a set of numbers or functions, is the justification for saying that each element of the ideal contains or is divisible by the ideal, since it possesses all the properties in question. Kronecker makes use of another image, in some respects simpler, viz., $a_1 u_1 + a_2 u_2 + ... + a_k u_k$ or $a_1 + a_2 u + ... + a_k u^{k-1}$, where $u, u_1, u_2, ..., u_k$ are indeterminates. This is not called an *ideal* because the term had already been appropriated by Dedekind with a different meaning, but it takes the place of Dedekind's ideal.

Thus at the outset we can form a natural conception of what should be meant by saying that an ideal $(a_1, a_2, ..., a_k)$ contains or is divisible by another $(\beta_1, \beta_2, ..., \beta_l)$. The conditions should be that each of $a_1, a_2, ..., a_k$ is an element of $(\beta_1, \beta_2, ..., \beta_l)$; for all elements of $(a_1, a_2, ..., a_k)$ will then possess all the properties possessed in common by all the elements of $(\beta_1, \beta_2, ..., \beta_l)$, and this apart from the fact that we may be unable to state explicitly what these properties are.

Again we can give a natural meaning to the G.C.M. and the L.C.M. of two ideals. The G.C.M. or H.C.F. of $(a_1, a_2, ..., a_k)$ and $(\beta_1, \beta_2, ..., \beta_l)$ should be an ideal $(\gamma_1, \gamma_2, ...)$ contained in both such that every ideal contained in both is contained in $(\gamma_1, \gamma_2, ...)$. There is one and only one such ideal, viz. the ideal $(a_1, a_2, ..., a_k, \beta_1, \beta_2, ..., \beta_l)$, cf. § 23. The L.C.M. should be an ideal $(\gamma_1, \gamma_2, ...)$ which contains both and such that every ideal which contains both contains $(\gamma_1, \gamma_2, ...)$. Again there is one and only one such ideal, viz. the ideal whose elements consist of all elements of $\omega(a)$ containing both $(a_1, a_2, ..., a_k)$ and $(\beta_1, \beta_2, ..., \beta_l)$. These elements constitute an ideal by definition.

But the crux lies in the difficulty of attaching a natural meaning to the term *product*. The product of two ideals should be an ideal whose properties consist of the product of the properties of the two ideals, and to this *product of properties* we cannot attach a meaning *a priori* from the definition of an ideal. Moreover the aggregate of the products of any element of $(a_1, a_2, ..., a_k)$ and any element of $(\beta_1, \beta_2, ..., \beta_l)$ does not constitute an ideal. The best that can be done is therefore to define the product of these two ideals to be the ideal $(..., a_i\beta_j, ...)$, $i=1, 2, ..., k, j=1, 2, ..., l$. This ideal includes all products $a\beta$ of elements of the two ideals, and in addition all sums of such products. It could not be told beforehand to what a theory based on so tentative a definition might lead.

We may say that the fact of an ideal containing another is a case of true divisibility if it always follows as a necessary consequence that the first is the product of the second and a third ideal (the converse being true by definition). This is exactly what Dedekind proved to be the case for all ideals of algebraic numbers and relatively whole algebraic functions of one variable, but only by means of a long series of subsidiary theorems. It followed that any such ideal could be uniquely expressed as a product of prime ideals. We know however that this is not true for ideals of functions of more than one variable, since it is not true for modules of rational functions. Also it is not true for ideals of absolutely whole algebraic functions of one variable ; e.g. (x) contains $(x, 2)$* but is not the product of $(x, 2)$ and a third ideal ; for the residual $(x)/(x, 2)$ is (x), and (x) is not the product of $(x, 2)$ and (x).

Kronecker's theory (Kr) concerns whole algebraic functions in general, and one of its remarkable features is that it applies to absolutely whole as well as to relatively whole functions. The absolute theory is based on the following fundamental theorem, which is proved by König (K, p. 78) :

If $f_1, f_2, ..., f_k$ *are any* k *polynomials in* $u_1, u_2, ...,$ Π *any product of coefficients of* $f_1, f_2, ..., f_k$ *taken one from each, and* $\Pi_1, \Pi_2, ...$ *the coefficients of the polynomial* $f_1 f_2 ... f_k$; *then* Π *satisfies identically an equation of the type*

$$\Pi^\rho + \Pi^{(1)} \Pi^{\rho-1} + \Pi^{(2)} \Pi^{\rho-2} + ... + \Pi^{(\rho)} = 0,$$

where $\Pi^{(i)}$ *is a homogeneous polynomial of degree* i $(i=1, 2, ..., \rho)$ *in* $\Pi_1, \Pi_2, ...$ *with rational integral coefficients.*

Kronecker gives the theorem in the second of the two memoirs referred to in (Kr), having discovered it after the first memoir was written. He states it for two polynomials f_1, f_2 in a single letter u or x. König gives the theorem in the more general form above. It is not generally necessary to introduce more than one letter or indeterminate u. If we suppose $f_1, f_2, ..., f_k$ to be polynomials of degrees $l_1, l_2, ..., l_k$ in a single letter u the number of the quantities Π is $(l_1 + 1)(l_2 + 1) ... (l_k + 1)$, while the number of the quantities $\Pi_1, \Pi_2, ...$ (which are sums of the quantities Π) is only $l_1 + l_2 + ... + l_k + 1$; and†

$$\rho = \frac{(l_1 + l_2 + ... + l_k)!}{l_1! \, l_2! \, ... \, l_k!}.$$

* In the relative theory $(x, 2) = (1)$, but not in the absolute theory. In the absolute theory a module in n variables can be of rank $n+1$ (cf. § 47); such in fact is any module which has some rational integer (but not unity) as a member, or any module which has no spread and is not (1). In both the absolute and relative theories the non-proper module (1) is without rank.

† The value found for ρ by König for the case $k=2$ is $(l_1 + l_2 + 1)!/l_1!(l_2 + 1)!$, which is not symmetrical in l_1, l_2. It can be proved that ρ need not be greater

Let the coefficients of $f_1, f_2, ..., f_k$ be absolutely or relatively whole algebraic functions of n variables $x_1, x_2, ..., x_n$. These all belong to and determine a corpus of functions. Let $M_1, M_2, ..., M_k$ be the ideals determined by the coefficients of $f_1, f_2, ..., f_k$ respectively, and M the ideal $(\Pi_1, \Pi_2, ...)$ determined by the coefficients of $f_1 f_2 ... f_k$. Then Π is any element of the basis of the ideal $M_1 M_2 ... M_k$, and $\Pi^{(i)}$ is an element of the ideal M^i (and of the involution M^i).

Kronecker says that a quantity Π which identically satisfies an equation of the above type (where $\Pi^{(i)}$ is any element of the ith power of a given ideal M) *contains M in the wider sense* of the word. Π contains M *in the strict sense* if it is an element of M, i.e. if it satisfies a *linear* identity of the above type. One ideal contains another (in the wider sense) if each element of (the basis of) the first contains the second (in the wider sense) ; and if each of two ideals M, M' contains the other (in the wider sense) M, M' are said to be equivalent in the wider sense. We denote such equivalence by $M \sim M'$, having already denoted strict equivalence by $M = M'$. Kronecker also remarks that (in the wider sense) if M contains M' and M' contains M'' then M contains M''. Consequently if $M \sim M'$ and $M \sim M''$ then $M' \sim M''$. *If $M, M_1, M_2, ..., M_k$ have the meanings given to them above we have* $M \sim M_1 M_2 ... M_k$.

This conception of wider equivalence is of considerable importance, and is specially applicable to Kronecker's theory. To any ideal M of a given corpus of functions there corresponds a unique *closed* equivalent ideal M_0 within the corpus. The elements of M_0 consist of all whole functions Π of the corpus which satisfy identically an equation of the type

$$\Pi^\rho + \Pi^{(1)} \Pi^{\rho-1} + \Pi^{(2)} \Pi^{\rho-2} + ... + \Pi^{(\rho)} = 0,$$

where $\Pi^{(i)}$ is an element of M^i. Any Π which satisfies identically an equation

$$\Pi^\sigma + \Pi_0^{(1)} \Pi^{\sigma-1} + \Pi_0^{(2)} \Pi^{\sigma-2} + ... + \Pi_0^{(\sigma)} = 0,$$

where $\Pi_0^{(i)}$ is an element of M_0^i, satisfies a linear identity of the same type and is an element of M_0. All ideals in the corpus equivalent to M are equivalent to M_0. A closed ideal may have relevant imbedded spreads; the closed module $(x_1^2, x_1 x_2)$ is an example.

If (speaking in the wider sense) M' contains M'' then MM' contains MM'', and conversely if MM' contains MM'' then M' contains M''; consequently if $MM' \sim MM''$ then $M' \sim M''$.

This theorem is not true for strict (or linear) equivalence, i.e. if $MM' = MM''$ it does not follow that $M' = M''$ (see § 24, Ex. i) unless M is an unmixed ideal of rank 1 (defined below).

than the smaller value given above, while for some of the products Π the value of ρ is less. In the cases of the first and last product Π it is evident that $\rho = 1$. If $l_1 = l_2 = 2$, ρ is 3 for the middle product Π, and 6 for the others, except the first and last.

Let $M = (a_1, a_2, ..., a_k)$, $M' = (a_1', a_2', ..., a'_{k'})$, $M'' = (a_1'', a_2'', ..., a''_{k''})$.

Then if M' contains M'' each element a' of the basis of M' satisfies an identity

$$a'^{\rho} + a^{(1)} a'^{\rho-1} + a^{(2)} a'^{\rho-2} + ... + a^{(\rho)} = 0,$$

where $a^{(i)}$ is an element of M''^i, i.e. a homogeneous polynomial in $a_1'', a_2'', ..., a''_{k''}$ of degree i with whole functions of the corpus for coefficients. Putting $a' a_j = a$ we have

$$a^{\rho} + a^{(1)} a_j a^{\rho-1} + a^{(2)} a_j^2 a^{\rho-2} + ... + a^{(\rho)} a_j^{\rho} = 0,$$

where $a^{(i)} a_j^i$ is an element of $(MM'')^i$; hence a contains MM'', i.e. MM' contains MM''. Conversely, given that MM' contains MM'', $a' a_j$ contains MM'' where a' is any element of M', i.e. we have an identity

$$(a' a_j)^{\rho_j} + \beta_j^{(1)} (a' a_j)^{\rho_j - 1} + \beta_j^{(2)} (a' a_j)^{\rho_j - 2} + ... + \beta_j^{(\rho_j)} = 0,$$

where $\beta_j^{(i)}$ is an element of $(MM'')^i = (a_1, a_2, ..., a_k)^i (a_1'', a_2'', ..., a''_{k''})^i$. Hence this identity is homogeneous and of degree ρ_j in $a_1, a_2, ..., a_k$, and arranging it in power products of these, each coefficient is homogeneous and of degree ρ_j in $a', a_1'', a_2'', ..., a''_{k''}$. There are k such equations for the same a', viz. when $j = 1, 2, ..., k$. The resultant of the k equations with respect to $a_1, a_2, ..., a_k$ is a homogeneous equation in $a', a_1'', a_2'', ..., a''_{k''}$ of degree $k \rho_1 \rho_2 ... \rho_k$, since it is homogeneous and of degree $\rho_1 \rho_2 ... \rho_k / \rho_j$ in the coefficients of the jth equation. Also since the resultant is homogeneous in $a', a_1'', a_2'', ..., a''_{k''}$, and is found by a purely algebraical process, we can find the coefficient of $a'^{k\rho_1\rho_2...\rho_k}$ in it by supposing all of $a_1'', a_2'', ..., a''_{k''}$ to be zeros. The resultant then becomes the resultant of $(a' a_1)^{\rho_1}, (a' a_2)^{\rho_2}, ..., (a' a_k)^{\rho_k}$, viz. $a'^{k\rho_1\rho_2...\rho_k}$. The coefficient of this term is therefore 1. Hence a' contains M'', i.e. M' contains M''.

The above properties are true not only for ideals but also for modules and for involutions whether of absolutely whole or relatively whole algebraic functions.

Kronecker's way of considering a set of whole algebraic quantities $a_0, a_1, ..., a_l$ (numbers or functions) is more direct than Dedekind's. He sets them in a frame or form*

$$a_u = a_0 + a_1 u + a_2 u^2 + ... + a_l u^l,$$

where u is an indeterminate. Instead of power products of one indeterminate u we could use l indeterminates $u_1, u_2, ..., u_l$ or power products of any less

---

* We use the term *form* here and later as meaning a *representation* which is not a function but is subject to algebraic laws and operations. The form becomes a function of u if u is regarded as a variable or parameter.

The notation a_u for a form is copied from König; and the notation for an ideal $M = (a_0, a_1, ..., a_l)$ is the same as that used in the text for a module. Kronecker's notation is quite different; e.g. he uses M for an element which is denoted above by a, and the term modular system as equivalent to divisor system or basis.

number. The indeterminates serve merely to separate the quantities $a_0, a_1, ..., a_l$. Kronecker then expands norm a_u in powers of u, viz.

$$\text{norm } a_u = (a_0 + a_1 u + ... + a_l u^l)(a_0' + a_1' u + ... + a_l' u^l)...$$
$$...(a_0^{(m-1)} + a_1^{(m-1)} u + ... + a_l^{(m-1)} u^l)$$
$$= F_0 + F_1 u + F_2 u^2 + ... + F_k u^k \quad (k = lm),$$

where $F_0, F_1, ..., F_k$ are whole rational functions of $x_1, x_2, ..., x_n$.

If $F_0, F_1, ..., F_k$ have an H.C.F. D, which may be a rational integer only, or a whole rational function of $x_1, x_2, ..., x_n$, then, having regard to the fact that norm a_u is the product of the above factors, we may say that $a_0, a_1, ..., a_l$ have something in common of the nature of a factor, which may be called their ideal common factor, and may be represented by the form a_u. So long as this factor D, which is the complete partial resolvent of rank 1 of the module $(F_0, F_1, ..., F_k)$, is only taken into account, while the partial resolvents of higher rank are neglected, Kronecker's theory is a theory of factorisation only.

Dedekind had established a theory of factorisation of whole algebraic numbers, which he subsequently extended to relatively whole algebraic functions of one variable. Considering that the factorisation of whole rational functions is exactly parallel to that of whole rational numbers the question naturally arises whether the factorisation of whole algebraic functions is parallel to that of whole algebraic numbers. Kronecker proved that it was absolutely parallel.

Kronecker says that a_u and norm a_u are *primitive* or *unit* forms if $D = 1$. This is legitimate in a theory of factorisation. Later he says that they are properly primitive only if the module* $(F_0, F_1, ..., F_k) = (1)$. If $D \neq 1$ then norm a_u/D is a unit form. Kronecker names $a_u/(\text{norm } a_u/D)$ an "algebraic modulus or divisor," which may be interpreted, an "equivalent of a_u" in respect to divisibility and factorisation. König names a_u/ϵ_u, where ϵ_u is any unit form in $\omega(a)$, an *ideal whole quantity* of $\omega(a)$; and accepts the rather absurd paradox that the sum of two such quantities is their H.C.F. It would be preferable to name a_u/ϵ_u an *ideal whole form*. He proves that such ideal forms can be uniquely resolved (in the sense of equivalence) into products of prime ideal forms, and shows how for a given form the prime factors can be actually found.

To compare two forms a_u and a_u' Kronecker considers the fraction a_u'/a_u and rationalizes the denominator by multiplying numerator and denominator by norm a_u/a_u, which is a (strictly) whole form in $\omega(a)$. If the new numerator a_u' norm a_u/a_u is divisible by the D of the new denominator norm a_u then the form a_u' is said to be divisible by the form a_u. If further the quotient of

---

* The module $(F_0, F_1, ..., F_k)$ is the aggregate of all whole rational functions $A_0 F_0 + A_1 F_1 + ... + A_k F_k$, and the ideal $(F_0, F_1, ..., F_k)$ in the domain $\omega(a)$ is the aggregate of all functions $\beta_0 F_0 + \beta_1 F_1 + ... + \beta_k F_k$, where $\beta_0, \beta_1, ..., \beta_k$ are elements of $\omega(a)$.

a_u' norm a_u/a_u by D is a unit form, then $a_u'/a_u = \epsilon_u'/\epsilon_u$, where ϵ_u, ϵ_u' are both unit forms, and the ideal forms a_u, a_u' are equivalent as regards divisibility. The divisibility of a_u' by a_u is the same thing as the divisibility of the ideal $(a_0', a_1', \ldots, a'_{l'})$ by the ideal $(a_0, a_1, \ldots, a_l)$ in the case of algebraic numbers and relatively whole algebraic functions of one variable; but not in other cases. This, as we show below, is a consequence of the fact that in these two cases $(F_0, F_1, \ldots, F_k) = (D)$.

Let M be the ideal $(a_0, a_1, \ldots, a_l)$, and, as before, let

$$\text{norm } a_u = F_0 + F_1 u + \ldots + F_k u^k \qquad (k = lm).$$

Then, in the Galoisian domain $\Omega(a, a', \ldots, a^{(m-1)})$, the ideal $(F_0, F_1, \ldots, F_k)$ is equivalent to the product of M and its conjugates M', M'', $\ldots$, $M^{(m-1)}$, by the fundamental theorem; and another ideal $(F_0', F_1', \ldots, F'_k)$ obtained in a similar way from any other basis of M is equivalent to $(F_0, F_1, \ldots, F_k)$, i.e. a homogeneous equation of degree ρ exists between $F_i', F_0, F_1, \ldots, F_k$, in which the coefficient of $F_i'^\rho$ is 1, and the other coefficients are whole elements of the Galoisian domain. By rationalizing the equation it follows that the *modules* $(F_0, F_1, \ldots, F_k)$, $(F_0', F_1', \ldots, F'_k)$ are equivalent. Hence we may define the rank of the ideal M and of the form a_u to be the rank of the module $(F_0, F_1, \ldots, F_k)$. We may also say that the ideal M is unmixed in the wider sense if the closed module equivalent to $(F_0, F_1, \ldots, F_k)$ is unmixed.

A *principal ideal* is an ideal (β) having a basis consisting of a single element β.

It can be proved without difficulty that *the only ideal in a given corpus $\Omega(a)$ equivalent to a principal ideal (β) is the ideal (β) itself.*

The ideal M above is called an *unmixed ideal of rank* 1 if $(F_0, F_1, \ldots, F_k)$ is a principal ideal, i.e. if $(F_0, F_1, \ldots, F_k) = (D) \neq (1)$.

Suppose now that $M = (a_0, a_1, \ldots, a_l)$ is an unmixed ideal of rank 1, and that the form a_u' is divisible by a_u in the sense defined above. Then, putting

$$\frac{\text{norm } a_u}{a_u} = \beta_0 + \beta_1 u + \ldots + \beta_{k-l} u^{k-l},$$

we are given that $(a_0' + a_1' u + \ldots + a'_{l'} u^{l'})(\beta_0 + \beta_1 u + \ldots + \beta_{k-l} u^{k-l})$ is divisible by D. Hence $a_i' \beta_j$ is divisible by D, i.e. $a_i' \beta_j = D \beta_{ij}$. Hence

$$a_i'(\beta_0 + \beta_1 u + \ldots + \beta_{k-l} u^{k-l}) = D(\beta_{i0} + \beta_{i1} u + \ldots + \beta_{i,k-l} u^{k-l}).$$

Multiplying by a_u, and putting $F_i = D\phi_i$, we have

$$a_i'(\phi_0 + \phi_1 u + \ldots + \phi_k u^k) = (\beta_{i0} + \beta_{i1} u + \ldots + \beta_{i,k-l} u^{k-l})(a_0 + a_1 u + \ldots + a_l u^l).$$

Hence $(\beta_{i0}, \beta_{i1}, \ldots, \beta_{i,k-l})(a_0, a_1, \ldots, a_l) \sim (a_i')(\phi_0, \phi_1, \ldots, \phi_k) = (a_i')$, since $(\phi_0, \phi_1, \ldots, \phi_k) = (1)$, and (a_i') is a principal ideal. Hence

$$(a_0', a_1', \ldots, a'_{l'}) = (\ldots, \beta_{ij}, \ldots)(a_0, a_1, \ldots, a_l) \quad (i = 0, 1, \ldots, l', j = 0, 1, \ldots, k-l).$$

Conversely, if an ideal $(a_0', a_1', \ldots, a'_{l'})$ contains an ideal $(a_0, a_1, \ldots, a_l)$ the form a_u' is divisible by the form a_u, since a_i norm a_u/a_u, and therefore also a_j' norm a_u/a_u, is divisible by D.

Hence an ideal $(a_0', a_1', \ldots, a'_{l'})$ *which contains an unmixed ideal* $(a_0, a_1, \ldots, a_l)$ *of rank 1 is the product of* $(a_0, a_1, \ldots, a_l)$ *and a third ideal* $(\ldots, \beta_{ij}, \ldots)$, *i.e. it has* $(a_0, a_1, \ldots, a_l)$ *as a true factor.* This includes Dedekind's principal result, since all ideals considered by him are unmixed ideals of rank 1. If the ideal $(a_0', a_1', \ldots, a'_{l'})$ is also unmixed the quotient $(\ldots, \beta_{ij}, \ldots)$ is also unmixed. Hence there exists a theory of factorisation for the whole aggregate of unmixed ideals of rank 1 in a corpus $\Omega(a)$.

It follows that an unmixed ideal $(a_0, a_1, \ldots, a_l)$ *of rank 1 can be multiplied by a second unmixed ideal of rank 1 so as to become the principal ideal* (β), *where* β *is any element of* $(a_0, a_1, \ldots, a_l)$.

If it be true that any ideal of rank 1 must contain an unmixed ideal of rank 1, which is obvious in the two cases considered by Dedekind, but has not been proved in general so far as I know, then *any unmixed ideal of rank 1 is a unique product of unmixed prime ideals of rank 1.* For, assuming the truth of the hypothesis, it can be shown that any two given unmixed ideals M, M' of rank 1 which have a common factor must have an H.C.F., viz. the unmixed ideal M'' of rank 1 such that $(M, M') = M'' M'''$, where M'''' is either (1) or of rank > 1. It can be easily proved that the ideals M'', M'''' thus defined are unique, and that any unmixed ideal of rank 1 which is a factor of M and of M' is a factor of M''; hence M'' is the H.C.F. of M and M'. In the cases considered by Dedekind (M, M') is itself an unmixed ideal of rank 1, and $M'' = (M, M')$. I cannot say whether this resolution into prime factors is exactly what is meant by Kronecker in his statement XIII, p. 89; and I cannot attach any true meaning to the parallel statement XIII°, p. 92, regarded as an extension of XIII.

Kronecker also considers another kind of divisibility of a form a_u' by a form a_u, which is more adaptable to the *general* theory of ideals. A form a_u' might be defined as divisible by a_u if the ideal $M' = (a_0', a_1', \ldots, a'_{l'})$ contains the ideal $M = (a_0, a_1, \ldots, a_l)$ in the strict sense. This definition is open to the objection that $a_u' \beta_u$ could be divisible by $a_u \beta_u$ without a_u' being divisible by a_u. The objection disappears when a wider definition is taken, viz. that a_u' is divisible by a_u if M' contains M in the wider sense.

The necessary and sufficient condition that any given ideal

$$M' = (a_0', a_1', \ldots, a'_{l'})$$

may contain any other given ideal $M = (a_0, a_1, \ldots, a_l)$ in the wider sense is that the ideal corresponding to a_u' norm a_u/a_u contains the ideal $(F_0, F_1, \ldots, F_k)$ corresponding to norm a_u in the wider sense, *which is the same thing* as containing the module $(F_0, F_1, \ldots, F_k)$ in the wider sense. In other words, it is necessary and sufficient that each of the $k - l + l' + 1$ coefficients a'' of the form a_u' norm a_u/a_u should satisfy identically an equation of some degree ρ which is homogeneous in a'', F_0, F_1, $\ldots$, F_k, the coefficient of a''^ρ being 1, and the other coefficients whole rational functions.

Printed in the United States
By Bookmasters